SWITCH-MODE POWER SUPPLY DESIGN

SWITCH-MODE POWER SUPPLY DESIGN

P.R.K. CHETTY

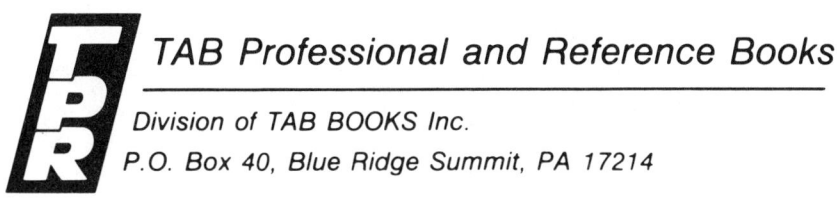

TAB Professional and Reference Books

Division of TAB BOOKS Inc.
P.O. Box 40, Blue Ridge Summit, PA 17214

Author makes no representation that the use of the material described herein will not infringe on existing or future patent rights, nor do the descriptions contained herein imply the granting of licenses to make, use, or sell equipment constructed in accordance therewith.

FIRST EDITION
FIRST PRINTING

Copyright © 1986 by TAB BOOKS Inc.
Printed in the United States of America

Reproduction or publication of the content in any manner, without express permission of the publisher, is prohibited. No liability is assumed with respect to the use of the information herein.

Library of Congress Cataloging in Publication Data

Chetty, P. R. K.
 Switch-mode power supply design.

Includes index.
 1. Electronic apparatus and appliances—Power supply. I. Title.
TK7868.P6C46   1986   621.381′044   85-27708
ISBN 0-8306-2631-X

Contents

	Preface	vii
	Introduction	ix
1	**Switch-Mode Power Supplies—an Introduction**	1
2	**Modeling and Analysis**	7

Current Injected Equivalent Circuit Approach to Modeling Switching dc-dc Converters 8
Current Injected Equivalent Circuit Approach to Modeling of Switching dc-dc Converters in Discontinuous Inductor Conduction Mode 17
CIECA: Application to Current Programmed Switching dc-dc Converters 26
Current Injected Equivalent Circuit Approach to Modeling and Analysis of Current Programmed Switching dc-dc Converters (Discontinuous Inductor Conduction Mode) 33
Modeling and Analysis of CUK Converter Using Current Injected Equivalent Circuit Approach 40

3	**Design and Measurements**	47

Modeling and Design of Switching Regulators 48
Closed Loops—on Track for Testing Switchers 60
Measurement of Magnitude and Phase of Switching Regulator Transfer Functions and Loop Gain 69

4	**Computer-Aided Design**	84

SPICE-2 CAD Package for the Design of Switching Regulators 84

5	**Practical Design Examples**	93

Design of a 2.8 kW Off-Line Switcher Using PWM Push-Pull Converter 94

Microprocessor-Controlled Digital Shunt Regulator 105
Multiphase Operation of Self-Oscillating Switching Regulator 119
Dc-dc Converter Maintains High Efficiency 123
Linear Power Supplies 127
Improvements to Power Supplies 128

6 ICs for Switch-Mode Power Supplies   131

Control ICs for Switch Mode Power Supplies 132
IC Timers as Controllers for Switch-Mode Power Supplies 140

7 Spacecraft Power Systems   145

Spacecraft Power Systems 146
Improved Power Conditioning Unit for Regulated Bus Spacecraft Power System 151

8 Reliability   159

Design for Reliability 160
Reliability and Redundancy 167
Reliability and Failure Mode and Effects Analysis 170

Index   177

Preface

In the 1960s, demands of the space programs led to the development of highly reliable, efficient and lightweight electrical power systems for spacecraft. Despite the limited supply of available energy onboard the spacecraft, engineers found innovative solutions for power processing and management of electrical power. These helped usher in the era of modern power electronics. Today, similar limitations on sources of available energy are becoming a prime design consideration in everyday electric power processing.

Power electronics is entirely devoted to switch-mode power conversion and deals with modern problems in analysis, design, and synthesis of electronic circuits as applied to efficient conversion, control, and regulation of electrical energy. Design and optimization of dc-to-dc converters, which offer the highest power efficiency, small size and weight, and high performance, are also included in power electronics.

These dc-to-dc converters with isolation transformers can have multiple outputs of various magnitudes and polarities. The regulated power supply of this type has wide applications, particularly in computer systems, wherein a low-voltage high-current power supply with low output ripple and fast transient response are mandatory. In addition, these converters connected in a particular configuration result in switched-mode ac power amplifiers with enough bandwidth and high efficiency. Off-line switchers, dc-to-dc single and multiple output power supplies, bi-directional power supplies (battery chargers and dischargers), dc-to-ac inverters, dc-to-ac uninterruptible power supplies, dc-to-ac motor control, power servo control, robotics, and switching audio amplifiers, etc., are some of the examples of switch-mode power supplies.

Switch-mode power supplies have come into widespread use in the last decade. An essential feature of efficient electronic power processing is the use of semiconductors in a power switching mode (to achieve low losses) to control the transfer of energy from source to load through the use of pulse-width modulated or resonant techniques. Inductive and capacitive energy storage elements are used to smooth the flow of energy while keeping losses to a low level. As the frequency of switching increases, the size of the magnetic and capacitive elements decreases in a direct proportion. Because of their superior performance, i.e., high efficiency, small size and weight and relatively low cost, they are displacing conventional linear (dissipative as they operate in linear or conduction mode) power supplies even at very low power levels.

The modeling, analysis, and design of these switching dc-to-dc converters have been extensively carried out, and it is commonly believed that the designs in the commercial use today employ the simplest possible converter topology for dc-to-dc conversion. Industry has been quick to realize that the energy saving technique also affords the opportunity to make dramatic reductions in equipment size and weight. Consumer and industry applications are expanding rapidly.

Among the various approaches developed for modeling and analysis of the switching dc-to-dc converters, the current-injected equivalent-circuit approach and state space average approach are used in producing a linear-equivalent circuit model that correctly represents the nonlinear converter properties for the static as well as dynamic ac small signal at low frequency levels, the essential features of the input and output transfer properties. Availability of the above model allows choice of the best converter for a specific application and optimization of the feedback loop of a regulator containing such a converter.

Also the models enable to design the switching regulators for stable operation with large bandwidth, fast transient response, and good line rejection. This is because the design can apply the standard method or circuit analysis applicable to linear feedback control systems using linear feedback control theory. However, the validity of these models and the design can only be made by measuring the frequency response of the system to check the accuracy of the loop gain and phase. Thus, in the design of the feedback system it is necessary to make measurement of the loop gain on the practical circuit as a function of the frequency to ensure that the circuit operates as analytically predicted, or get feedback from the measurement to correct the analytical prediction.

Thus, switch-mode power supplies have received considerable attention because of their high performance features and it is my intention to make available all the work I carried out in power electronics (part of which has already been published in the conference proceedings, journals and well received), so that the power electronics community may enjoy its fruits. Thus, power electronics is covered in detail in this book.

Introduction

This book is an assembly/collection of the papers I have published (some of them are coauthored) and papers specially written for this book. Most of the papers were presented in conferences and in professional journals. Conference papers were published only in the conference proceedings, which are largely obscure and unavailable, consequently I thought that it would be of great benefit for the power electronics community to publish them as a book. This volume contains papers dealing with modeling, analysis, design, measurement including computer aided design, power supply practical design examples, and reliability aspects. It comprises 23 papers arranged in a systematic order into eight chapters in such a way it is easy to understand.

Chapter 1 gives an introduction to switch-mode power supplies. The evolution of the power electronics is presented first. Then it stresses the importance of power systems in all electronic equipment and systems as it is the main source of power with which each and every electronic equipment works, and the successful operation of all electronic equipment critically depends upon proper and reliable functioning of the power system. The second paper presents a definition of power system, classification of power supplies, brief operational description of dissipative (linear) power systems and nondissipative (switch-mode) power systems, as an introduction to mathematical modeling and analysis of switch-mode power supplies.

Chapter 2, entitled *Modeling and Analysis*, contains five papers. In this, a new current-injected equivalent circuit approach (CIECA) to modeling switching dc-dc converter power stages is developed, which starts with the current-injected approach and results in either a set of equations that completely describe input properties or an equivalent circuit model valid at small-signal low-frequency levels. As an example, the duty-ratio programmed converter operating in continuous inductor conduction (CIC) mode is modeled.

The same current-injected equivalent circuit approach is employed to model the converters operating in the following modes, (i) duty ratio programmed converters operating in discontinuous inductor conduction (DIC) mode; (ii) current programmed converters operating in continuous inductor current (CIC) mode; and (iii) current programmed converters operating in discontinuous inductor conduction (DIC) mode.

As the Cuk converter has the merits of both buck and boost converters, namely, it has non-pulsating input and output currents and it can buck and boost the input voltage to result in the required output voltage, it is worthwhile to have the linear equivalent circuit model for the Cuk converter. Also it is an optimum topology converter as it uses the minimum number of components. Hence, the mathematical modeling and analysis of the Cuk converter is also included.

Having known the models for switching dc-dc converters, now the design of regulators using above mentioned basic or extended converters is appropriate and is thus dealt in Chapter 3. Various building blocks of a switching regulator are described in detail and mathematical models are developed for all building blocks in terms of transfer functions that enable one to design a switching regulator for stability, desirable bandwidth, line rejection, and transient response. A step-by-step procedure to design compensation is illustrated using two examples. Various networks for compensation and their transfer functions are presented.

Having designed compensation and implemented in circuitry, only the measurements will confirm the accuracy of the modeling and design, which is dealt with in the second and third papers. Different techniques to measure the magnitude and phase of switching regulator transfer functions and loop gain are presented. Now to accelerate the design process a computer-aided design approach for switching regulators is presented in Chapter 4.

Chapter 5 includes six papers to give more insight into the practical hardware design aspects of switch-mode power supplies. In the first paper, design of an Off-Line 2.8 kW Switcher employing Pulse-Width Modulated Push-Pull Converter as the power stage is presented including the step-by-step design procedure that eliminates the trial and error approach and results in fewer man-hours spent in development. In the second paper, a new approach to the design of power systems is presented in which a microprocessor is used as a controller for a digital shunt regulator (DSR). As the microprocessor need not be dedicated to the DSR, it can simultaneously be used for battery management and for charge regulator and or discharge regulator control.

Because of the power handling limitations of the qualified semiconductor devices, such elements are quite often used in parallel to meet the large power requirements. In such a situation the power is shared inphase between the power handling devices since all of them switch simultaneously. This simultaneous switching action creates problems in filtering and electromagnetic screening. Multiphase operation is employed primarily to minimize these problems. The multiphase operation of the self-oscillating switching regulator is reported in the third paper. In the fourth paper, a dc-dc converter is analyzed, which has to supply a constant power load even when the input voltage undergoes large variations and it is shown that the efficiency of the converter is lower at high voltages than at low voltages. A new base drive is implemented that improved the efficiency at high voltages.

In most dc input regulated power supplies, regulation is poor when the desired output voltage is less than the source internal reference voltage. In addition, circuit considerations usually limit the minimum reference voltage attainable and consequently the minimum regulated output voltage possible. The circuit presented in the fifth paper has a configuration that can bring the reference voltage virtually to zero, and overcomes both of the problems.

Simple changes like adding a component appropriately can improve the performance of a power supply. Two such improvements are presented as examples, one for a linear power supply and the second for a switch-mode power supply in the sixth paper.

Chapter 6 entitled *ICs for switch-mode power supplies,* presents the ICs for control and other aspects of power supply design, the use of which enhanced the growth as well as the benefits of switch-mode power supplies. The ICs offer the advantages of compactness, accuracy, reproducibility, higher performance through reduction of parasitics, and the economies of mass production. The first paper presents ICs for control, protection and instrumentation of free-running as well as driven-type power supplies.

Not only the regulator ICs or pulse-width modulating regulator ICs are used for switch-mode power supplies, but other ICs are used as well. This is illustrated in the second paper, where an integrated-circuit timer is used as the control element for a simple dc-dc converter regulator with current foldback, the current step-up converter regulator and a polarity reversing voltage step-up converter regulator.

Chapter 7 includes two papers that deal with spacecraft power systems. The first paper describes a typical spacecraft power system consisting of solar cells, storage batteries, and power conditioning and control electronics. The main differences between power systems for low earth orbit satellites and geo-synchronous orbit satellites are also presented. In the second paper, an improved power conditioning unit developed for regulated bus spacecraft power systems based upon the principle of using a common control block for charge, discharge, and shunt regulators is presented. In addition heavy elements like inductor for charge and discharge regulators and output capacitors for shunt and discharge regulators are made common for the integrated system.

Chapter 8 deals with the reliability aspects of the power systems. Reliability is the main requirement of any equipment or system. Unreliability can mean lots of waste. The important ways of improving the reliability of any electronic system and various methods to be followed in designing the systems for high design reliability compared to the reliability due to components, manufacturing techniques, etc., are covered in the first paper.

The reliability and redundancy aspects are examined with special reference to the power supplies. Though the inherent circuit reliability could be maximized by circuit design, judicious selection of components, etc., the chance failure that could partially or totally jeopardize a mission can be taken care of only by adopting redundancy to assure the overall mission reliability at the required level. The second paper describes different approaches to redundancy such as standby redundancy, load sharing redundancy, majority logic redundancy, partial redundancy and shared mode of standby redundancy. High performance, high levels of reliability and lower costs are the primary considerations in the design of power systems. A reliability analysis is presented in the third paper, which provides a measure of reliability designed into the system. Also included is the failure mode and effects analysis (FMEA) whose purpose is to identify and eliminate, where possible, critical single point failures.

Chapter 1

Switch-Mode Power Supplies—an Introduction

SWITCH-MODE POWER SUPPLIES

1.0 Introduction

A power system basically processes the power to convert it from one form (input) to another form (required output). The hardware used to carry out this power processing is known as a power supply, regulator, dc-dc converter, battery eliminator, etc., and this field is known as *Power Processing Electronics* or simply *Power Electronics*.

The power supply is one of the important elements of any piece of electrical or electronic equipment. This is because it provides power to energize all the electrical or electronic circuits and makes the equipment operate. The successful operation of any piece of electrical or electronic equipment depends upon proper and reliable functioning of the power supply. The specifications of the power supply are closely linked with the equipment. The stringent demands on performance, weight, volume, reliability and cost make the design of the power supply, a truly challenging exercise. In general, the power supply shall be able to deliver the regulated power at specific voltage and current levels meeting the equipment requirements. This can be anywhere from a fraction of a watt to a few thousands of watts and few volts to thousands of volts and the voltage can be dc or ac. In the case of ac output voltage, the frequency can be anywhere from a few cycles to few thousands of cycles. It is very essential to have very high efficiency. Also heatsinking and forced air requirements increase with lower efficiency and the system becomes bulkier and heavier, which is not acceptable. Thus, the weight and volume of the power supply has to be as small as possible. It is not only desirable but also essential to have the cost of the power supply to be as low as possible.

In view of the fact that the mean time between failure (MTBF) of any piece of equipment is closely linked with that of the power supply, the design and technology of the power system have received a great deal of attention. The last decade has witnessed significant advances in Power Electronics resulting in the development of reliable, lightweight and high efficiency power systems with large MTBF.

The energy source for most of the equipment is ac. The other types of energy sources are batteries, solar cells, etc. Thus, the sources of energy are available in the form of ac or dc. However, most of the electrical and electronic equipment operates with dc voltage. Within the equipment some portion may work at a different dc voltage than the other portion. Some equipment may require more than one dc voltage for internal operation.

Few pieces of equipment work on ac voltage directly. The power system converts the voltage from one type to another and from one level to another level and in the case of ac from one frequency to another frequency. Thus, the power system processes the power while matching the impedance of the energy sources to that of the loads or equipment. In other words, the power system conditions the outputs of the energy sources so as to match with the requirements of the various loads or equipment.

2.0 Definitions

As explained above, the primary function of any power supply is to provide a predetermined constant output voltage when the input voltage and/or output current vary widely and there is a possibility of change in operating temperature. The degree to which a power supply provides a constant output voltage under the above conditions is the basic figure of merit of the power supply. Accordingly, regulation is defined as given below.

$$\text{Line Regulation}(\%) = [\Delta E_o / \Delta E_{in}]/E_o \times 100$$

where ΔE_o = Change in output voltage
ΔE_{in} = Change in input voltage
E_o = Nominal output voltage

$$\text{Load Regulation}(\%) = [E_{nl} - E_{fl}]/E_o \times 100$$

where E_{nl} = Output voltage with no load
E_{fl} = Output voltage with full load

E_o = Nominal output voltage

Temperature Coefficient =

$\pm [E_{o\,max} - E_{o\,min}]/[E_o(T_{max} - T_{min})] \times 100$

where $E_{o\,max}$ = Output voltage at max. rated temperature (T_{max})
$E_{o\,min}$ = Output voltage at min. rated temperature (T_{min})
T_{max} = Maximum operating temperature
T_{min} = Minimum operating temperature

3.0 Classification of Power Systems

The power systems employ different approaches to process the power and to convert it from one form to other. The power systems can be classified as (i) *Dissipative (Linear) Power Systems* and (ii) *Nondissipative (Switch-Mode) Power Systems.* As the name indicates the dissipative systems are those which dissipate more in the conversion process, thereby operate inefficiently requiring large heatsink area. The dissipation varies as a function of input voltage and load fluctuations and hence they result in poor efficiency. However, these power systems exhibit low EMI and less ripple characteristics. These power systems can be further classified into series and shunt types.

The nondissipative power systems operate in the switch mode resulting in high efficiencies. Nondissipative power systems employ some kind of highly efficient dc-to-ac conversion process. Once the dc power is converted into ac power, it is easy to level-up or level-down the voltage by employing low weight and high frequency transformers. In most of the practical applications, transformer isolation is very essential and it is a must for equipment operated by human beings. Final dc voltage can be obtained by rectifying and filtering the ac waveform. There are two approaches to dc-to-ac conversion. The first approach is the square wave or pulse-width modulated (PWM) approach where the dc input is chopped at a high frequency rate to result in a square wave. This square wave voltage can be levelled-up or levelled-down by using transformers. The output can be rectified and filtered to get dc at a different voltage level than the input. The duty ratio determines the amplitude of the output voltage if the input voltage is constant. The duty ratio of the square wave can be varied to obtain regulation if there are variations in input voltage or load. The second approach is the resonant mode approach wherein dc input is applied through controlled switches to a LC resonant circuit. In this approach, the rate of energy exchange is not governed by an independent clock but by the resonant frequency of the energy storage elements.

Thus power supplies (dc-dc converters) can be viewed as a linear or nonlinear LC or non-LC oscillator(square or sine wave) coupled through a transformer-rectifier, to a low-pass filter. Fig. 1 shows the power supply classification tree.

3.1 Dissipative or Linear Power Sytems

In the case of linear type, the pass transistor is operated in active conduction mode such that the voltage across the pass transistor is always equal to the difference between the input voltage and output voltage. Diode voltage and emitter follower regulators are the simplest regulators one can conceive. These are normally employed only for coarse regulation, with low output current requirements, and where efficiency is not an important consideration.

3.1.1 Series Regulator

The most widely used of all linear regulator circuits is the series regulator, the basic configuration of which is shown in Fig. 2. There is a series control device usually a transistor in either the common-emitter or common-collector configuration that acts on a signal from the control circuit or error amplifier and prevents the output voltage from fluctuating. The control circuit may take many forms, but it will always have some sort of reference with which it compares the received sample of the output voltage and amplifies the difference. The resulting error signal corrects the drive of the se-

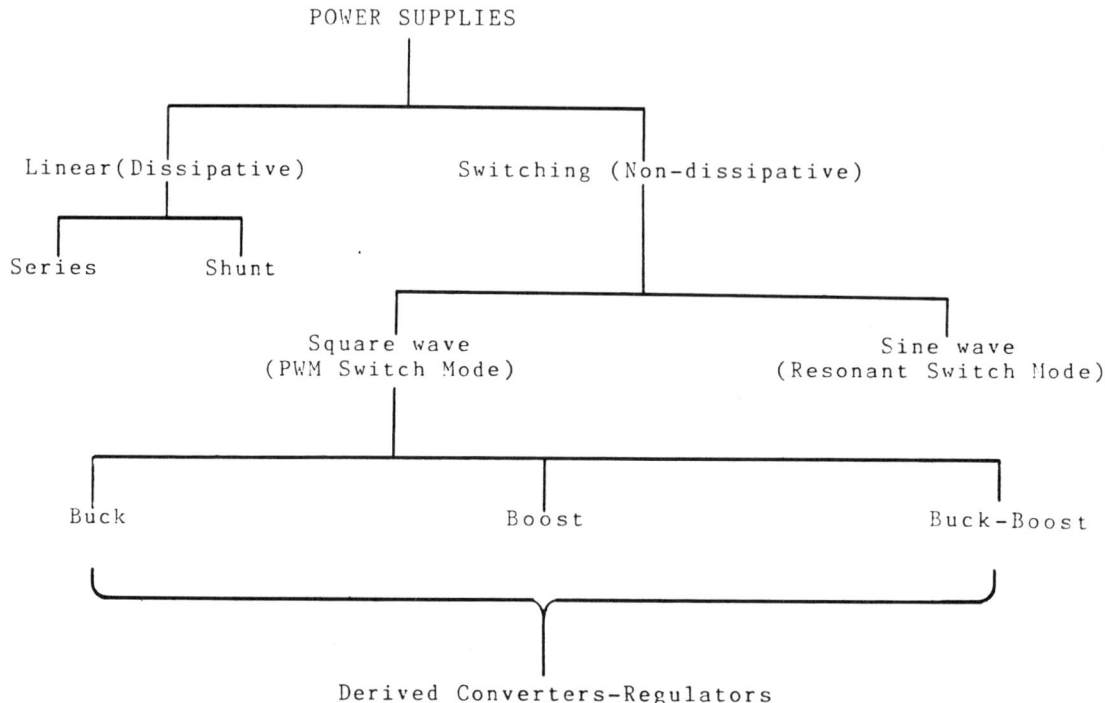

Fig. 1. Power supply classification tree.

ries transistor, so that the collector-emitter voltage is always the difference between the input voltage and the desired output voltage, even if there are variations in the input voltage and/or output load. As in this regulator, the series-pass transistor always operates in the active region, the dissipation is high. Hence, the series transistor must be chosen carefully to avoid the second breakdown. This regulator is generally preferable for high voltage, medium current loads.

3.1.2 Shunt Regulator

The basic configuration of a shunt regulator is shown in Fig. 3. Here the shunt control element (transistor) must be capable of withstanding the entire output voltage, but it does not have to carry the full load current unless required to regulate from full load to no load. Since the series-dropping resistor used with the shunt regulator has high dissipation, the efficiency of this regulator is poor. This regulator is preferable for medium to low voltages and high output currents with relatively constant loads.

3.2 Nondissipative Power Systems

As explained above, in the case of the switch-

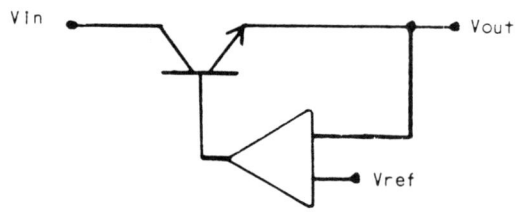

Fig. 2. Basic configuration of a series regulator.

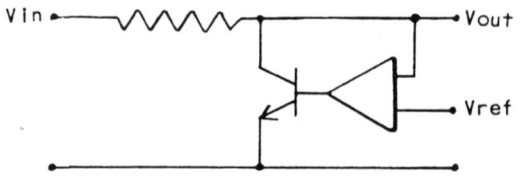

Fig. 3. Basic configuration of a shunt regulator.

ing or nondissipative power systems, the pass transistor operates in ON (saturation) or OFF (cut-off) mode and hence the dissipation in the pass transistor is minimum. The regulation is achieved by controlling the duty ratio of the pass transistor. The output voltage, in these regulators, can be greater than, equal to, or less than the input voltage. The regulators can be divided into three types, viz., (i) buck type, (ii) boost type and (iii) buck-boost type.

3.2.1 Buck Regulator

In a buck regulator, the output voltage is always less than the input voltage and can be practically anywhere between 10% and 90% of the input voltage. This means that a switching regulator can be used as a dc step-down transformer with highest efficiency. Figure 4 shows a simple buck converter power stage. The output voltage is compared with a stable reference voltage and the amplified error signal is used to generate a pulse-width modulated waveform, which controls the switch ON/OFF periods. When the switch is turned-ON, current flows through the inductor and into output capacitor and the load. When the output voltage exceeds the reference voltage, the switch is turned-OFF. At this instant, the stored energy in the inductor reverses its polarity, takes the path through the diode and sends the current into the load while the voltage is maintained by the capacitor. When all the stored energy in the inductor is used up, the capacitor discharges and the output voltage decreases. At this step, the switch is turned-ON and the process continues such that the output voltage is maintained very close to the reference voltage.

3.2.2 Boost Regulator

In a boost regulator, the output voltage is always higher than the input voltage. A schematic of a boost power stage is shown in Fig. 5. When the switch is turned-ON, the current flows through the inductor and energy is stored in it. When the switch is turned-OFF, the stored energy in the inductor tends to collapse and its polarity changes such that it adds to the input voltage. Thus, the voltage across the inductor and the input voltage are in series and together charge the output capacitor to a voltage higher than the input voltage.

3.2.3 Buckboost Regulator

This regulator operates on the principle of both buck and boost. When the switch is turned-OFF, the inductor releases the stored energy similar to an automobile ignition system. The output voltage is, of course, determined by the rate of discharge of the inductor. Rapid discharge results in a lower voltage and vice versa. Figure 6 shows a simple buckboost power stage. When the switch is ON, the inductor charges and stores energy. When the switch is OFF, the stored energy tries to collapse reversing its polarity, thus sending current into the output capacitor and load. This takes output voltage to an opposite polarity to that of the input voltage.

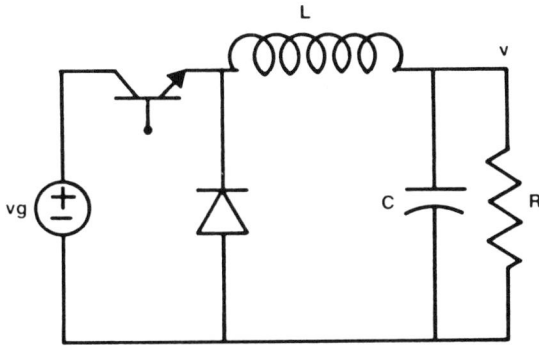

Fig. 4. Buck converter power stage.

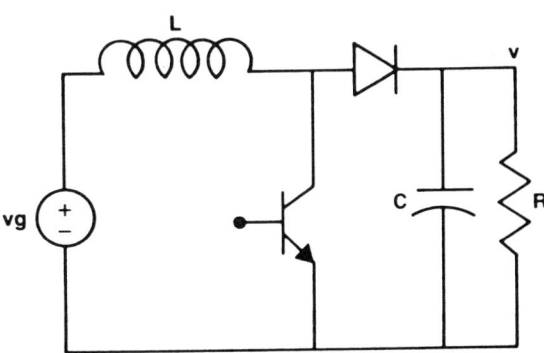

Fig. 5. Boost converter power stage.

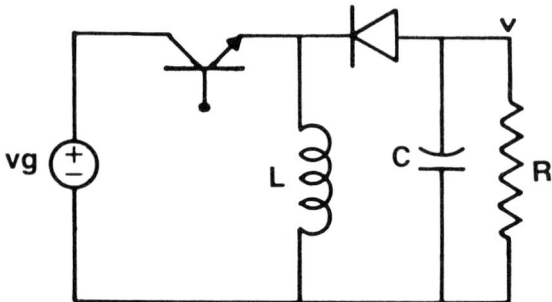

Fig. 6. Buckboost converter power stage.

3.2.4 Other Types of Converter-Regulators

The above described regulators are the three basic converter-regulators. All other converters like, Cuk, Bell Lab, Weinburg, Vanable, etc., are derived from these basic converter-regulators.

3.2.5 Free Running and Driven Types

The switching regulators can be of the free-running type or driven type. As is clear from above description, the output voltage is compared with a reference voltage and the amplified error voltage is used to generate a PWM signal, which is used to drive the regulator switch such that the output voltage is maintained at a predetermined level. This PWM signal can have one of the following characteristics, (i) Fixed ON period and variable OFF period; (ii) Variable ON and OFF periods with fixed frequency; (iii) fixed OFF period and variable ON period. For a specific use, a trade off between the various characteristic features has to be carried out depending upon the requirements and an appropriate configuration has to be selected.

To enhance the power capability, the above regulators are usually connected in parallel and are operated in phase-shift (multiphase) mode to reduce the EMI and problems of electromagnetic screening. To improve the performance characteristics such as regulation, transient response, line rejection, etc., the feedback control loop of the above regulators is modified to sense the ac changes besides the dc changes on the output voltage and the changes in the input voltages.

Chapter 2

Modeling and Analysis

Current Injected Equivalent Circuit Approach to Modeling Switching dc-dc Converters 8

Current Injected Equivalent Circuit Approach to Modeling of Switching dc-dc Converters in Discontinuous Inductor Conduction Mode 17

CIECA: Application to Current Programmed Switching Dc-Dc Converters 26

Current Injected Equivalent Circuit Approach to Modeling and Analysis of Current Programmed Switching dc-dc Converters (Discontinuous Inductor Conduction Mode) 33

Modeling and Analysis of CUK Converter Using Current Injected Equivalent Circuit Approach 40

CURRENT INJECTED EQUIVALENT CIRCUIT APPROACH TO MODELING SWITCHING DC-DC CONVERTERS

A new current injected equivalent circuit approach (CIECA) to modeling switching dc-dc converter power stages is developed, which starts with the current injected approach and results in either a set of equations which completely describe input and output properties or an equivalent linear circuit model valid at small signal, low frequency levels.

This approach to modeling switching dc-dc converter power stages has the merits but not the demerits of both the electronic equivalent circuit state space average approach and the current injected control type approach, namely, 1) the modeling is very clear and is simple whether the converter operates in continuous or discontinuous inductor conduction modes, 2) the modeling results in an equivalent circuit which is very close to the actual converter, and 3) the equivalent circuit can be used directly in the computer for theoretical predictions like SPICE, etc.

1.0 Introduction

Modeling switching converters has received considerable attention in recent years and a number of methods have been developed ranging from analytic to design oriented, and the results range from specific numeric solutions to general equivalent circuit models. A good review of these approaches is attempted in [1,2]. Of the various approaches to modeling switching converters existing to date only the electronic equivalent circuit state space average approach [3-6], and the current injected control type approach [2,7,8,9] are well received. Whatever the approach used to get the converter transfer properties, the result is of course the same, however one approach gives more information about the converter properties compared with the other. The current injected approach represents control type techniques which arrive at a block diagram linearized description of the nonlinear system and models only transfer properties. The electronic equivalent circuit state space average approach models input and output in addition to transfer properties. The equivalent circuit approach might be preferred by electronic circuit designers, and those accustomed to the control type might prefer the current injected control type approach to modeling switching converters.

It is thought that one of the most useful benefits of the electronic equivalent circuit state space average approach is the ease with which more complicated converter structures can be analyzed. The equivalent circuit also leads to the physical insight that permits optimum design. But a thorough study and application of both approaches reveal that 1) the current injected control type approach in continuous inductor conduction (CIC) mode is equally as easy compared with equivalent circuit state space average approach, and 2) the electronic equivalent circuit state space average approach is not clear [4] or becomes more complex and cumbersome [6] in discontinuous inductor conduction (DIC) mode, whereas the current injected control type approach is very clear and becomes easier. These two facts led me to favor the current injected control type approach to model input and output, as well as the fact that transfer properties simultaneously develop an equivalent circuit using the current injected control type approach. Thus this approach is called the current injected equivalent circuit approach (CIECA). This modified approach has the merits of the above mentioned two approaches, the most important of which are the following.

1) Linear equivalent circuit is developed to give physical insight into the converter circuit that permits optimum design.
2) The availability of the input model avoids the fresh start in the analysis of cascaded converters.
3) Input and output as well as transfer properties are modeled.
4) The analysis is clearer and easier in the DIC mode.

The development of the CIECA to modeling switching converters is presented in detail. To

---

©1981 IEEE. Reprinted with permission from *IEEE Transactions on Aerospace and Electronic Systems*, Vol. AES-17, No. 6, pp. 802-808, Nov. 1981.

demonstrate this approach, modeling and analysis is carried out for the basic three converters: buck, boost, and buckboost. Section 2 contains the detailed development of modeling switching converter power stages using the CIECA. The CIECA is demonstrated in Section 3 by applying it to the boost converter power stage. Following the same approach, the modeling is carried out for buck and buckboost converters and the results are presented in Section 4. Section 5 compares the merits of the CIECA with those of the electronic equivalent circuit state space average approach and the current injected control type approach. Section 6 concludes this new approach, the CIECA, to modeling switching dc-dc converter power stages.

2.0 CIECA

The development of the CIECA allows a unified treatment of a large variety of converter power stages, since this approach is very simple and easy to apply. Physical reasoning in each step of this modeling is included and the modeling is attempted for the basic three converters: buck, boost, and buckboost. The end result of modeling is either a set of equations representing the transfer properties of converter, or a linear equivalent circuit model for the nonlinear converter.

The following conventions and notations are followed in the modeling and analysis: $d_1 T_s$ is the interval during which the transistor is turned on and the diode is off, $d_2 T_s$ is the interval during which the transistor is turned off and the diode is on, $d_1 + d_2 = 1$, and $T_s = 1/f_s$ is the switching period. Capitalized quantities are used for steady-state values and quantities with carets are used for small perturbations.

The CIECA is outlined in the flowchart of Fig. 1, which is very general and applicable to various power stages. The first step in this process is to identify the nonlinear part of the converter circuit and to linearize only that part of the converter as the remaining part is inherently linear. Thus the converter power stage is identified as both nonlinear and linear and in fact the nonlinear part of the converter determines the average current injected into the linear part of the converter. Hence this approach becomes simple as the linearization is achieved by averaging the current through the nonlinear part that is injected into the linear part of the converter. One of the two parts contains the switch and is supposed to be the nonlinear part as it takes different connections depending upon whether the switch is turned on or off. The second part does not contain any switch, remains the same throughout the switching period, and is inherently linear.

Figure 2 contains the source voltage v_g, a three-terminal block, and a parallel RC network. Assume the three-terminal block simply contains an ideal switch as shown in Fig. 2 (B); theoretically the circuit results in a converter though there are practical limitations. Initially, as the capacitor is not charged, a very large current will flow through the switch, only limited by the equivalent series resistance (ESR) of the capacitor. This large current can damage the switch itself. In addition the ripple will be very large as the capacitor has to supply current to the load during the period when the switch is turned off. An inductor is added to alleviate these problems, as shown in Fig. 2 (C). Now the current through the three-terminal block is limited and controlled by the value of the inductor L which is of course affected by the parallel RC network. This is true whether the 2 connects to the 1 (to the source) or to the 3 (to the ground). Thus the buck converter would have resulted. Now the clockwise rotation of Fig. 2 (C) results in Fig. 2 (D). We know that this circuit is a buckboost converter. A further step in clockwise rotation of Fig. 2 (D) results in Fig. 2 (E). Again we are familiar with this circuit, the boost converter. This is one of the views which explains the development of basic converters. In fact this is how the nonlinear (three-terminal block) and linear (parallel R and C) parts are identified in basic converters. A similar method can be followed for any switching converter power stage.

As mentioned above, the three-terminal block represents a nonlinear part whereas the second (parallel R and C) part represents a linear part. Also one can see that the three-terminal block injects current into the second part resulting in appropriate

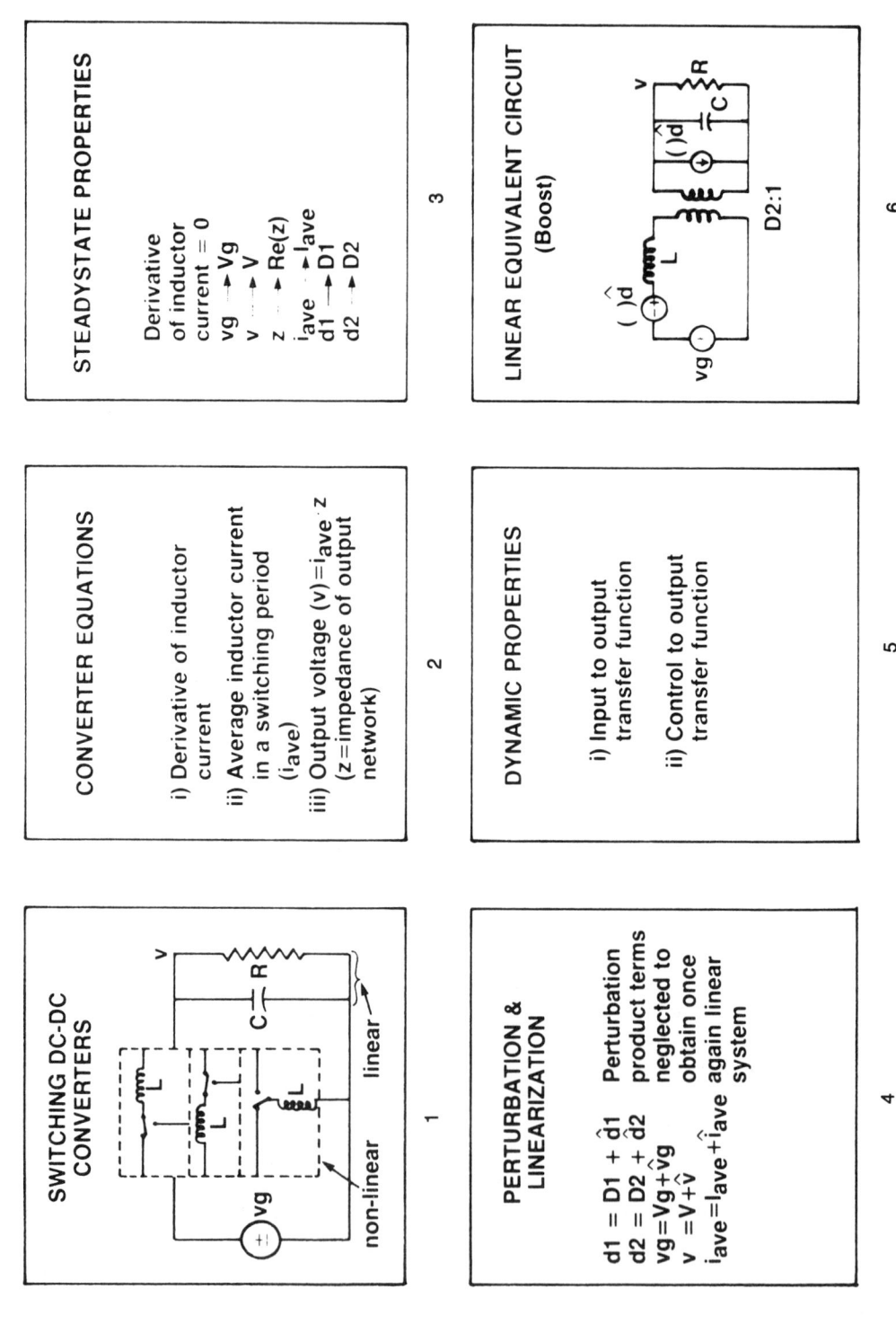

Fig. 1. Flowchart of CIECA to modeling switching dc-dc converters in duty ratio programmed CIC mode.

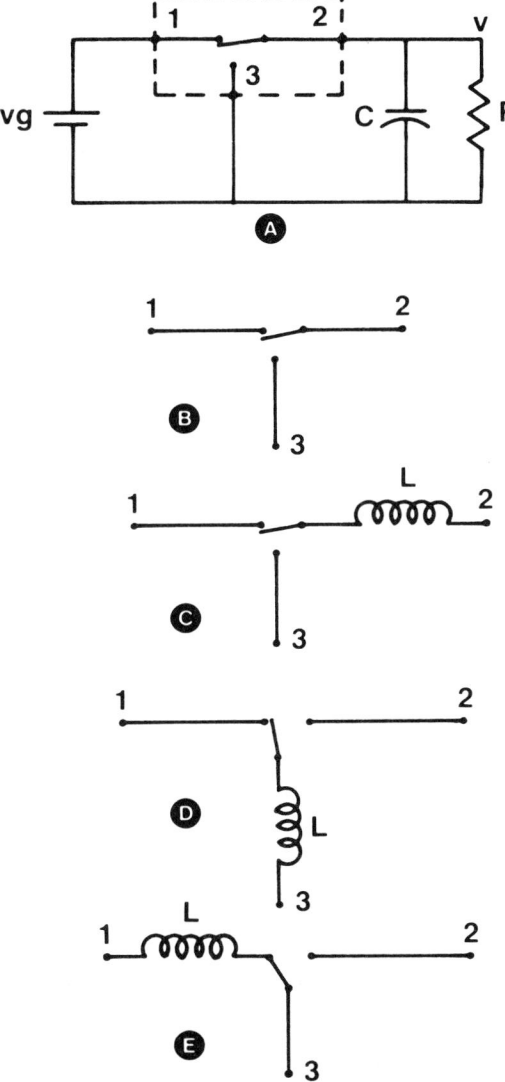

Fig. 2. (A) Probable basic circuit with which switching converters would have developed. (B) Three terminal block shown separately. (C) Inductor is added to (B) at an appropriate place. (D) Clockwise rotation of (C). (E) Clockwise rotation of (D).

output voltage. The following are a set of relationships referring to the converter diagram and current and voltage waveforms shown in Fig. 3:

1) average current (i_{ave}) determined by the first part, injected into the second part in a switching period;

2) derivative of the inductor current function of the value of the inductor, the voltage across that in each subinterval in a switching period;
3) relationship between average injected current and output voltage $v = (i_{ave})(z)$, where z is the impedance of the linear part of the converter.

Now a steady-state solution is achieved by setting derivatives and perturbations to zero (Fig. 1, box 3). Since the converter equations in (Fig. 1, box 2), are linear, superposition holds and can be perturbed (Fig. 1, box 4) by the introduction of a small ac variation over the steady-state operating point. As we know, the independent driving inputs are vg and d, the perturbation in these two inputs causes the perturbation in i and v. The small ac variation from the steady-state operating point is negligible compared with the steady-state operating point values, i.e., $\hat{v}/V, \hat{v}g/Vg, \hat{d}/D, \hat{i}/I$ (each) $\ll 1$.

Using the above approximations, nonlinear second-order terms are neglected to obtain once again a linear set of equations. Now only the ac part is retained which describes the small signal, low fre-

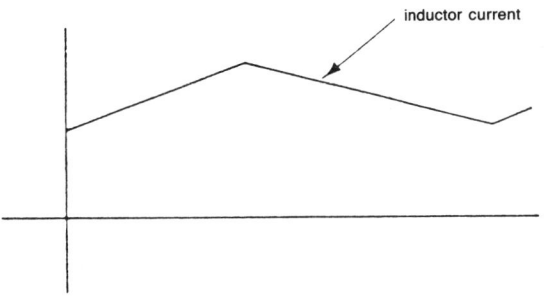

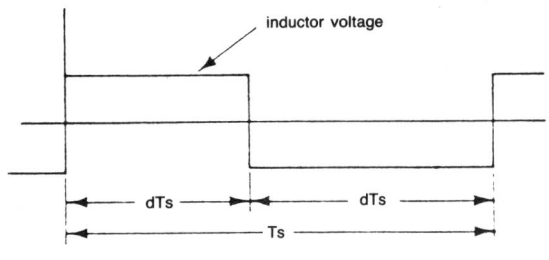

Fig. 3. Typical inductor current and voltage waveforms in buck converter.

quency behavior of the converter. Using this set of equations, the input-to-output and control-to-output transfer functions (Fig. 1, box 5) are written. Using the same set of equations an equivalent circuit (Fig. 1, box 6) is drawn which represents the input and output small signal, low frequency properties of the nonlinear converter.

Although the outlined method follows in terms of equations and arrives at an equivalent linear circuit model, one can proceed from (Fig. 1, box 2) in a parallel way using equivalent circuit models. As in the first method, a perturbation and linearization are carried out and from the resulting circuit models a final linear equivalent circuit model is obtained similar to that of (Fig. 1, box 6).

Even though both paths have identical results, one need use only one method depending on his taste; however the circuit model path gives more physical insight into the qualitative nature of the results, especially the right half-plane zeros in boost and buckboost converters. Thus the CIECA to modeling switching dc-dc converter power stages derives the linear equivalent circuit which completely describes the input and output small signal, low frequency properties of the nonlinear converter power stage, in addition to the transfer properties.

3.0 Boost Converter Modeling

We now demonstrate the method for the boost converter power stage shown in Fig. 4. The switches are assumed to be ideal. Parasitics and storage time effects of the transistor switch are not included for simplicity. The CIECA can be applied to the converters operating in both CIC and DIC modes whether they are duty ratio programmed or current programmed. Similarly the CIECA can be applied whether the converter operates in free-running or in fixed-frequency mode. However, the present modeling is limited to fixed-frequency duty ratio programmed converters operating in CIC mode.

Inductor current and voltage waveforms for the boost converter are shown in Fig. 5. The shaded portion shows the amount of current injected into the output linear circuit (parallel R and C) and the interval during which the current injected is $d2\,Ts$.

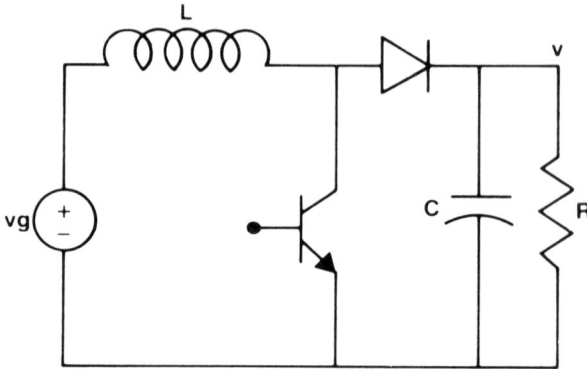

Fig. 4. Boost converter with all parasitics and storage time effects neglected.

The average inductor current injected into the output circuit during a switching period is given by

$$i_{ave} = i\,(d2) \qquad (1)$$

where i is the inductor current. The derivative of the inductor current is given by

$$L(di/dt) = [vg - v\,(d2)] \qquad (2)$$

Therefore the output voltage is

$$v = [i_{ave}\,R]/[1 + sRC] \qquad (3)$$

where $R/(1 + sRC)$ is the impedance of the output network. The steady-state conditions can now be found by using (1)-(3) and setting the derivative to zero. Therefore the above equations reduce to

$$V/Vg = 1/D2; \quad I = V/[R(D2)]. \qquad (4)$$

Equations (1)-(3) are perturbed around the steady-state operating point, and second-order nonlinear terms are neglected once again to obtain the linear small signal model

$$\hat{i}_{ave} = \hat{i}(D2) - I\hat{d}$$
$$L(d\hat{i}/dt) = \hat{v}g - \hat{v}(D2) + V\hat{d} \qquad (5)$$
$$\hat{v} = [R/(1 + sRC)]\,\hat{i}_{ave}$$

The input-to-output and the control-to-output

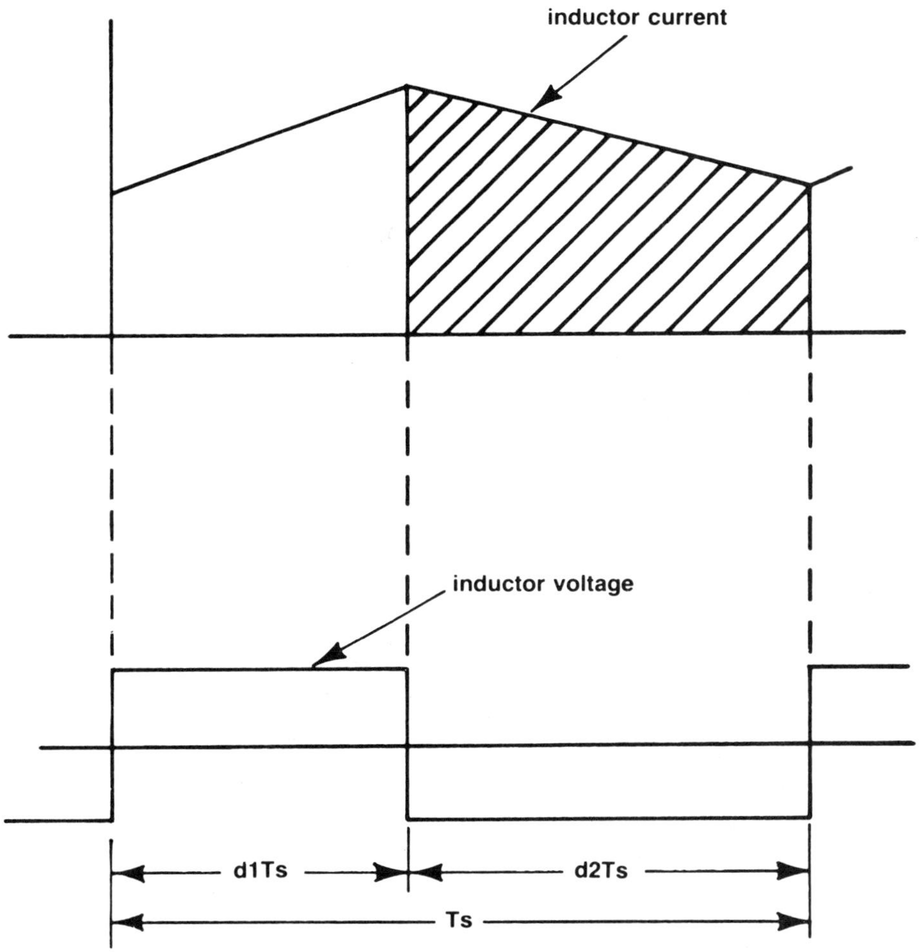

Fig. 5. Inductor current and voltage waveforms of the boost converter in Fig. 4.

transfer functions are obtained from (5) by first taking Laplace transform

$$\hat{v}(s)/\hat{v}g(s) = (1/D2)[1/(1 + SL/RD^2_2 + S^2 LC/D^2_2)]$$
$$\hat{v}(s)/\hat{d}(s) = (Vg/D_2) (1 - SL/RD^2_2)/ \quad (6)$$
$$(1 + SL/RD^2_2 + S^2LC/D^2_2)$$

These transfer functions are the same as those obtained by [3,6] using the electronic equivalent circuit state space average approach or the current injected control type approach. Using (5), an equivalent circuit as shown in Fig. 6 can be drawn. The dependent current and voltage generators are replaced by an equivalent transformer as shown in Fig. 7. This equivalent circuit is identical to the equivalent circuit model obtained using the electronic equivalent circuit state space average approach.

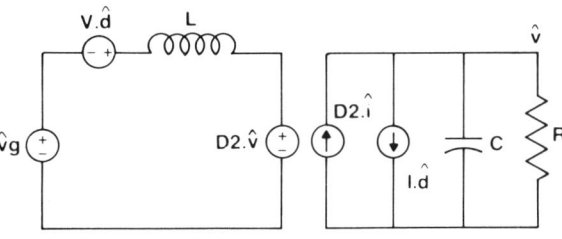

Fig. 6. Small signal low frequency linear equivalent circuit model of the boost converter of Fig. 4. This circuit can be directly used in computer simulation.

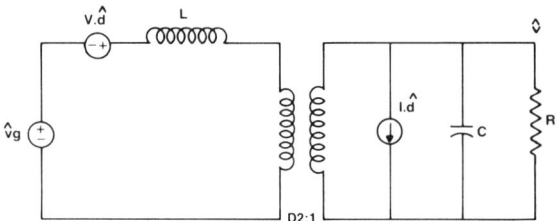

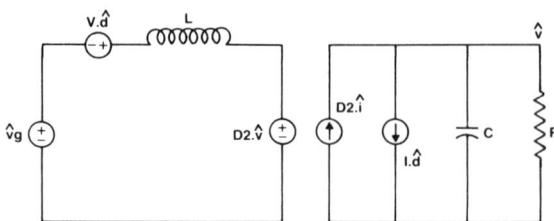

Fig. 7. Small signal low frequency linear equivalent circuit model of the boost converter replacing the dependent current and voltage generators by an equivalent transformer.

Fig. 9. Small signal low frequency linear equivalent circuit for boost converter of Fig. 4.

From the circuit of Fig. 7 one can see that there is a current generator (in place of a switch in the actual circuit of Fig. 4) between the L and the C. This current generator is moved to the input to put the circuit in a form that enables one to see that there is really a low-pass LC filter. Thus the movement of the current generator to input produces a frequency-dependent voltage generator. The sign of this generator is such that in the control-to-output transfer function, this results in a right half-plane zero. Also notice that the effective value of L now depends upon the steady-state duty ratio.

Hybrid Approach. A hybrid approach to modeling is demonstrated below which uses the equivalent circuit model immediately after the linearization of the nonlinear part of the converter. Using (1)-(3), an equivalent circuit as shown in Fig. 8 can be drawn. From this circuit we can see that during steady-state, the dc model is simply obtained by short circuiting the inductor and open circuiting the capacitor; then vg, v, $d1$, and $d2$ assume the steady-state values. Now the circuit is perturbed and second-order terms are neglected to once again obtain the linear system. One can see that Fig. 9 is the same as Fig. 6 and one will get an equivalent circuit as in Fig. 7 after the dependent generators are replaced with a transformer.

4.0 Modeling Buck and Buckboost Converters

The same method as described in Section 3.0 has been followed to model buck and buckboost converters. The converter and its equivalent circuit diagrams are shown in Fig. 10 for the buck converter, and shown in Fig. 11 for the buckboost. The results for the two transfer functions of principal interest, the input-to-output transfer function, and the control-to-output transfer function are as follows.

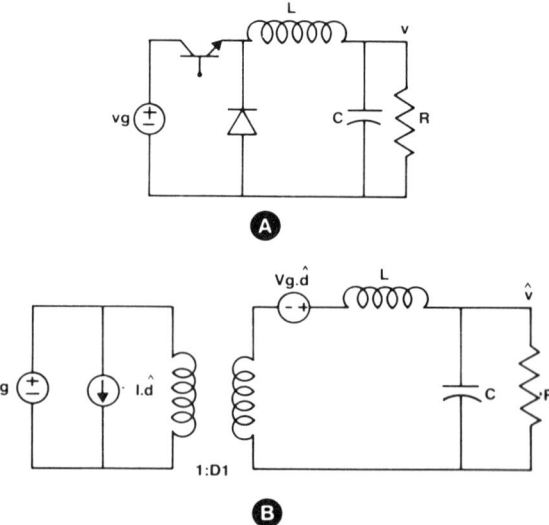

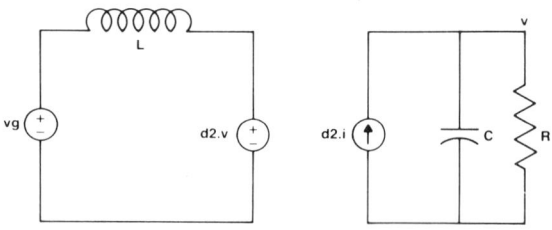

Fig. 8. Equivalent circuit for (1)-(3).

Fig. 10. (A) Buck converter. (B) Linear equivalent circuit model.

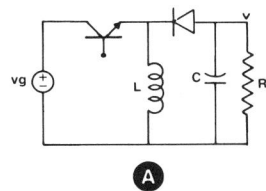

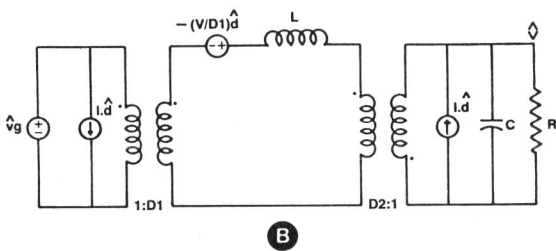

Fig. 11. (A) Buckboost converter. (B) Linear equivalent circuit model.

For the buck converter,

$\hat{v}(s)/\hat{v}g(s) = (D1)[1/(1 + SL/R + S^2 LC)]$
$\hat{v}(s)/\hat{d}(s) = (Vg) [1/(1 + SL/R + S^2 LC)]$.

For the buckboost converter,

$\hat{v}(s)/\hat{v}g(s) = (D1/D2) [1/(1 + SL/RD_2^2 + S^2 LC/D_2^2)]$
$\hat{v}(s)/\hat{d}(s) = (V/D1D2) (1 - SLD1/RD_2^2)/$
$(1 + SL/RD_2^2 + S^2 LC/D_2^2)$

The equivalent circuit diagrams and transfer functions obtained here for buck and buckboost are the same as those obtained in [3].

5.0 Comparison

Modeling and analysis for the duty ratio programmed buck, boost, and buckboost converters operating in fixed-frequency CIC mode are carried out and their linear equivalent circuit models developed using the new CIECA. The transfer functions are the same as those obtained in the current injected control type approach or in the electronic equivalent circuit state space average approach. In addition the equivalent circuit models developed are the same as those obtained in the state space average approach. Table I gives a detailed comparison of CIECA with the current injected control type approach and the electronic equivalent circuit state space average approach.

6.0 Conclusions

A new CIECA to modeling of switching dc-dc converters is developed and presented which describes the small signal, low frequency input-to-output and control-to-output transfer properties as well as input and output properties of the converter. To demonstrate this approach, the modeling is carried out for buck, boost, and buckboost converters operating in fixed-frequency, duty ratio programmed CIC mode. The results of the modeling and analysis are compared with those results obtained by using the current injected control type approach as well as by using the electronic equivalent circuit state space average approach.

Table 1 gives a detailed comparison of CIECA with two other approaches. It is very clear that the CIECA has the merits of both the other approaches. The merits of the CIECA will be substantiated by work to be published in which the modeling and analysis is carried out for the converters operating in the DIC mode, for the converters operating in current programmed mode, and for cascaded converters. This approach also allows power system designers with control background to use their control knowledge and still get equivalent circuit models to give more physical insight into the converter operation, thereby enabling them to make better designs. This approach is also encouraging and attractive to power system designers with circuit background as the modeling becomes simpler and is very clear in DIC mode.

References

1. Middlebrook, R.D., and Ćuk, S. (1977) Modeling and analysis methods for dc-dc switching converters. Presented at the IEEE International Semiconductor Power Converter Conference, Orlando, FL, 1977.
2. Fossard, A.J., and Clique, M. (1976) Modelisation des cellules elementaires. Technical Report 1, Contract 2590/75 AK, European Space Research and Technology Organization, Noordwijk, The Netherlands, 1976.

Table 1. Comparison of the CIECA to Modeling Switching Dc-Dc Converters with the Current Injected Control Type Approach and the Electronic Equivalent Circuit State Space Average Approach.

Parameter	Current Injected Current Type Approach	Electronic Equivalent Circuit State Space Average Approach	CIECA
Input and output equivalent circuit models	not available	available	available
Transfer functions	easier to obtain	easier to obtain	easier to obtain
Resemblance of equivalent circuit/block diagrams to actual converter	control type block diagram is too far away from actual converter	canonical models do have resemblance to actual converter	equivalent circuit models are much closer to physical converter circuit and the canonical models can be obtained after manipulations.
Modeling in CIC	easy and simple	easy	easy and simple
Modeling in DIC	easy and simpler	either not clear [4] or complex and cumbersome [6]	easier, simpler, and very clear
Physical reasoning for right half-plane zeros	not available	available	available
Use of block diagrams/ equivalent circuit models in computers for theoretical predictions	usable in analog and hybrid computers	[1] Cannot be used directly in computers, like using SPICE, etc.	can be used directly in computers, like using SPICE, etc.
Modeling approach	linear and nonlinear parts of the converter power stage are identified and only that part is linearized; hence this approach is simple and easy	complete converter is treated for linearization though only some part of the converter is nonlinear; hence this approach becomes complex	linear and nonlinear parts of the converter power stage are identified and only that part is linearized; hence this approach is simple and easy

[1] This does not mean that the canonical models are not useful. In fact, they are the models that for the first time helped in understanding that all the basic converters effectively have L-C filter properties though it is not seen in the actual converter circuits, especially in buckboost and boost where there is a switch between L and C.

3. Middlebrook, R.D., and Ćuk, S. (1976) A general unified approach to modeling switching converter power stages. In *IEEE Power Electronics Specialists Conference Record*, 1976, pp. 18-34.
4. Ćuk, S., and Middlebrook, R.D. (1977) A general unified approach to modeling switching dc-to-dc converters in discontinuous conduction mode. In *IEEE Power Electronics Specialists Conference Record*, 1977, pp. 36-57.
5. Hsu, S.-P., Brown, A., Rensink, L., and Middlebrook, R.D. (1979) Modeling and analysis of switching dc-dc converters in constant frequency current programmed mode. In *IEEE Power Electronics Specialists Conference Record*, 1979, pp. 284-301.
6. Brown, A. (1979) State space analysis of pulse width modulated switching converters. Technical Note T-58, Power Electronics Group, California Institute of Technology, Pasadena, Feb. 6, 1979.
7. Fossard, A.J., and Clique, M. (1979) Modeling and design of dc-dc converters using modern control theory, part 1: modelisation; part 2: open loop analysis and control design. Presented at the 3rd European Space Research and Technology Organization Spacecraft Power Conditioning Seminar, Noordwijk, The Netherlands, 1977.
8. Ferrante, J.G., Capel, A., Fossard, A.J., and Clique, M. (1977) A general linear continuous model for design of power conditioning units at fixed and free running frequency, In *IEEE Power Electronics Specialists Conference Record*, 1977, pp. 113-124.
9. Prajoux, R., Marpinard, J.C., and Jalade, J.

(1966) Establishment de modeles mathematiques pour regulaterus de puissance a modulation de largeur d impulsions (pwm); 2: models continus. *ESA Scientific and Technical Review*, 1966, 2 (1), 115-129.

CURRENT INJECTED EQUIVALENT CIRCUIT APPROACH TO MODELING OF SWITCHING DC-DC CONVERTERS IN DISCONTINUOUS INDUCTOR CONDUCTION MODE

A new current injected equivalent circuit approach (CIECA) to modeling switching dc-dc converter power stages is developed, which starts with current injected approach, and results in a set of equations which describe completely input and output properties and an equivalent linear circuit model valid at small signal low frequency levels.

This approach to modeling switching dc-dc converter power stages has the merits of two known approaches: i) electronic equivalent circuit state space average approach, ii) current injected control type approach, namely; a) the modeling is very clear and is simple whether the converter operates in continuous or discontinuous inductor conduction modes, b) results in an equivalent circuit which is very close to the actual converter, c) the equivalent circuit can be used directly in computer for theoretical predictions like SPICE, etc., d) devoid of the demerits of both the approaches mentioned.

Having developed and demonstrated for the converters operating in continuous inductor conduction modes[10], the CIECA is now extended to the converters operating in discontinuous inductor conduction mode.

1.0 Introduction

The modeling of switching converters has received considerable attention in recent years and a number of methods have been developed, ranging from analytic to design oriented, and the results range from specific numeric solutions to general equivalent circuit models. A good review of these approaches is attempted in [1], [2]. Among the various approaches to modeling switching converters existed to date, only the following two approaches are well received, namely: i) electronic equivalent circuit state space average approach [3-6]; ii) current injected control type approach [2], [7-9]. Whatever the approach is used to get the converter transfer properties, the result is, of course, the same; but one approach gives additional information about the converter properties compared to the other. The current injected approach on one hand, represents control type techniques which arrive at a block diagram linearized description of the nonlinear system and models only transfer properties; on the other hand, electronic equivalent circuit state space average approach models input and output in addition to transfer properties. The equivalent circuit approach might be preferred by electronic circuit designers and those accustomed to the control type might prefer current injected control type approach to modeling of switching converters.

In addition to the above comments, it is thought that one of the most useful benefits of electronic equivalent circuit state space average approach is the ease with which more complicated converter structures can be analyzed and the equivalent circuit leads to the physical insight that permits optimum design. But a thorough study and application of both approaches reveal much more interesting facts that a) the current injected control type approach in continuous inductor conduction (CIC) mode is equally easier compared to equivalent circuit state space average approach; b) electronic equivalent circuit state space average approach is not clear [4] or becomes more complex and cumbersome [6] in discontinuous inductor conduction (DIC) mode, whereas current injected control type approach is very clear and becomes more easy.

Because of these two facts, current injected control type approach is used to model input and output, as well as transfer properties simultaneously developing an equivalent circuit. Thus, this approach is hereafter called current injected equivalent circuit approach (CIECA).

Having developed and demonstrated the

© 1982 IEEE. Reprinted with permission from *IEEE Transactions on Industrial Electronics*, Vol. IE-29, No. 3, pp. 230-234, Aug. 1982.

switching dc-dc converters operating in continuous inductor conduction mode [10], the current injected equivalent circuit approach is now extended to the converters operating in discontinuous inductor conduction mode and presented in this paper. To demonstrate this approach, the modeling and analysis is carried out for the basic three converters, i.e., buck, boost, buckboost. The section following the introduction contains the detailed development of modeling of switching converter power stages using current injected equivalent circuit approach. This method is demonstrated by applying to the boost converter power stage in Section 3. Following the same approach, the modeling is carried (Appendix) out for buck, and buckboost converters and the results are presented in Section 4. Final section presents the conclusion.

2.0 Current Injected Equivalent Circuit Approach (CIECA)

The following conventions and notations are followed in the modeling and analysis:

$d_1 T_s$ — the interval during which the transistor is turned on and the diode is off,

$d_2 T_s$ — the interval during which the transistor is turned off and the diode is on,

$$d_1 T_s + d_2 T_s + d_3 T_s = T_s$$

$T_s = 1/f_s$ switching period.

The capitalized quantities are used for steady-state

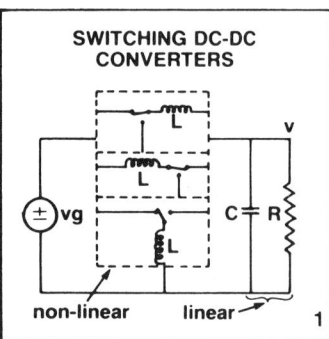

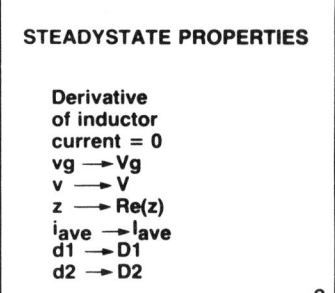

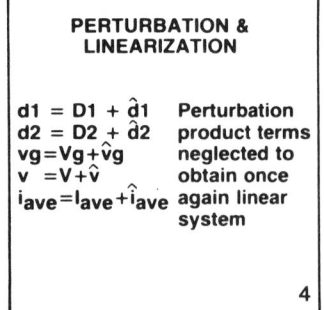

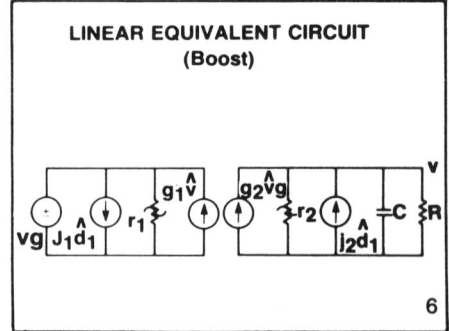

Fig. 1. Flowchart of current injected equivalent circuit approach to modeling switching dc-dc converters in the duty ratio programmed discontinuous inductor conduction mode.

values and the quantities with hats for the small perturbations.

The current injected equivalent circuit approach to modeling converters operating in discontinuous inductor conduction mode is outlined in the flowchart of Fig. 1, which is very general, applicable to various power stages. The first step in this process is to identify the nonlinear and linear parts of the converter circuit and linearize only the nonlinear part of the converter as the remaining part of the converter is inherently linear (Box 1). The nonlinear part of the converter determines the average current injected into the linear part. Now (Box 2), a set of relationships are written referring to the converter diagram and current and voltage waveforms shown in Fig. 2:

i) volt second balance on the inductor;
ii) average current (i_{ave}) injected into the linear part in a switching period;
iii) relationship between average injected current and output voltage $v = (i_{ave}) \times (z)$ where z is the impedance of the linear part of the converter.

Now steady-state solution is achieved by setting derivatives and perturbations to zero (Box 3). Since the converter equations in Box 2 are linear, superposition holds and can be perturbed (Box 4) by the introduction of a small ac variation over the steadystate operating point. As we know the independent driving inputs are vg and d, the perturbation in these two inputs cause the perturbation in i and v. Now making the small signal approximation,

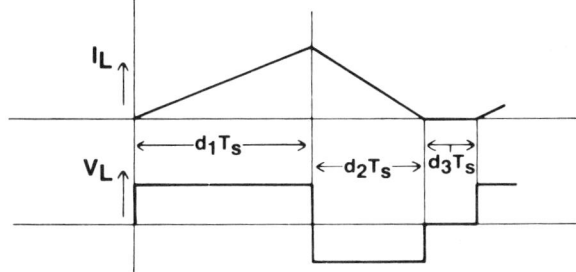

Fig. 2. Typical inductor current and voltage waveforms in buck converter.

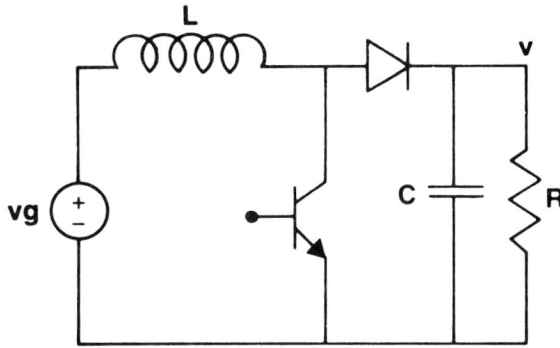

Fig. 3. Boost converter with all parasitics and storage time effects neglected.

namely, the small ac variation from the steadystate operating point are negligible compared to the steadystate operating point values, i.e., $\hat{v}/V$, $\hat{v}_g/V_g$, $\hat{d}_1/D_1$, $\hat{d}_2/D_2$, $\hat{i}/I$ (each) $\ll 1$. Using the above approximations, nonlinear second order terms are neglected to obtain once again a linear set of equations. Now only the ac part is retained which describes the small signal low frequency behavior of the converter. Using these sets of equations, the input to output and control to output transfer functions (Box 5) are written. Using the same set of equations, an equivalent circuit (Box 6) is drawn which represents the input and output small signal low frequency properties of the nonlinear converter.

Although the outlined method follows in terms of equations and arrives at the end an equivalent linear circuit model, one can proceed from Box 2 in a parallel way using equivalent circuit models. As in the first method, perturbation and linearization are carried out and from the resulted circuit models a final linear equivalent circuit model is obtained similar to that of Box 6. Both the paths result in identical results.

3.0 Boost Converter-Modeling

We now demonstrate the method for the boost converter power stage shown in Fig. 3. The switches are assumed to be ideal, and the present modeling is limited to fixed frequency duty ratio programmed converters operating in discontinuous

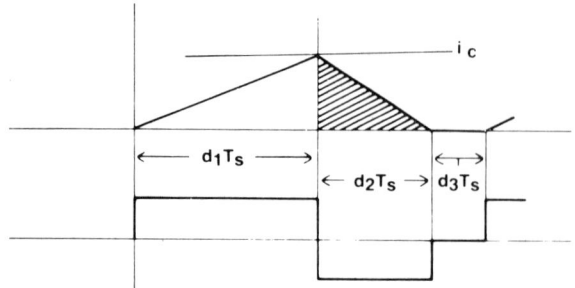

Fig. 4. Inductor current and voltage waveforms of the boost converter in Fig. 3.

inductor conduction mode. The RC constant is assumed to be much greater than the switching period T_s.

Inductor current and voltage waveforms for the boost converter are shown in Fig. 4. The shaded portion shows the amount of current injected into the output linear circuit (parallel R and C) and the interval during which the current injected is $d_2 T_s$. The average inductor current injected into the output circuit during a switching period is given by

$$i_{ave} = \frac{V_g \cdot d_1 \cdot T_s \cdot d_2}{2L}. \quad (1)$$

Volt second balance on the inductor

$$V_g(d_1 + d_2) = V \cdot d_2. \quad (2)$$

The output voltage is

$$v = (i_{ave} \cdot R)/(1 + sRC) \quad (3)$$

where $R/(1 + sRC)$ is the impedance of the output network. The steady-state conditions can now be found by using (1)-(3) and setting the derivative to zero, V_g, V, d_1, d_2, etc., assume steady-state values. Therefore, the above equations reduce to

$$V/V_g = 1 + \frac{D_1}{D_2} \quad \frac{V}{R} = \frac{V_g \cdot D_1 \cdot D_2 \cdot T_s}{2L}. \quad (4)$$

Equations (1)-(3) are perturbed around the steady-state operating point and second order nonlinear terms are neglected once again to obtain the linear small signal model

$$\hat{i}_{ave} = \frac{T_s}{2L}(\hat{V}_g \cdot D_1 \cdot D_2 + V_g \cdot \hat{d}_1 \cdot D_2 + V_g \cdot D_1 \cdot \hat{d}_2)$$

$$\hat{V}/V_g - V \cdot \hat{V}_g/V_g^2 = \hat{d}_1/D_2 - D_1 \cdot \hat{d}_2/D_2^2$$

$$\hat{V} = (R/(1 + sRC))\hat{i}_{ave}. \quad (5)$$

The input-to-output and the control-to-output transfer functions are obtained from (5) by first taking Laplace transform

$$\frac{\hat{V}}{\hat{V}_g} = (M) \frac{1}{1 + S/W_p}$$

$$\frac{\hat{V}}{\hat{d}} = \frac{2V}{2M - 1} \left(\sqrt{\frac{M-1}{KM}} \frac{1}{1 + S/W_p} \right)$$

where

$$M = \frac{V}{V_g} = \frac{D_1 + D_2}{D_2}$$

$$K = \frac{2L}{R \cdot T_s}$$

$$W_p = \left(\frac{2M - 1}{M - 1}\right) \frac{1}{RC}.$$

These transfer functions are the same as those obtained using electronic equivalent circuit state space average approach or current injected control type approach. Using (5) an equivalent circuit is drawn as shown in Fig. 5. This equivalent circuit is identical to the equivalent circuit model obtained using electronic equivalent circuit state space average approach.

A. Hybrid Approach. As mentioned in Section 2.0, a hybrid approach to modeling is

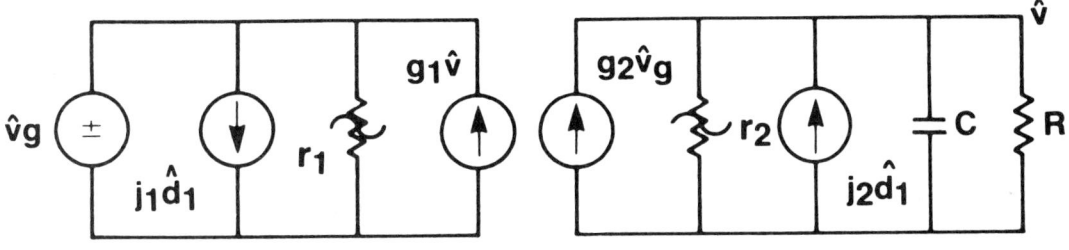

$$J_1 = \frac{2V}{R}\sqrt{\frac{M}{K(M-1)}} \qquad r_1 = \left(\frac{M-1}{M^3}\right)R \qquad g_1 = \left(\frac{M}{M-1}\right)\frac{1}{R}$$

$$J_2 = \frac{2V}{R}\left(\sqrt{\frac{1}{KM(M-1)}}\right) \qquad r_2 = \left(\frac{M-1}{M}\right)R \qquad g_2 = \frac{M(2M-1)}{M-1}\frac{1}{R}$$

Fig. 5. Small signal low frequency linear equivalent circuit for Boost converter of Fig. 3. This circuit can be directly usable in computer simulations.

demonstrated below, which uses the equivalent circuit model immediately after the linearization of the nonlinear part of the converter. Using (1)-(3), equivalent circuit, Fig. 6 can be drawn. From this circuit we can see that during steady-state, the dc model is simply obtained by open circuiting the capacitor and assuming steady-state values for vg, v, d_1, d_2. Now the circuit is perturbed, and second order terms are neglected to obtain the linear system once again. This is shown in Fig. 7 and is the same as Fig. 5.

4.0 Modeling of Buck and Buckboost Converters

The same method as described in the previous section has been followed (Appendix) to model buck and buckboost converters. The converter and its equivalent circuit diagrams are shown in Fig. 8 for the buck converter and in Fig. 9 for the buckboost converter. The results for the two transfer functions of principal interest, the input to output transfer function and the control to output transfer function are as follows.

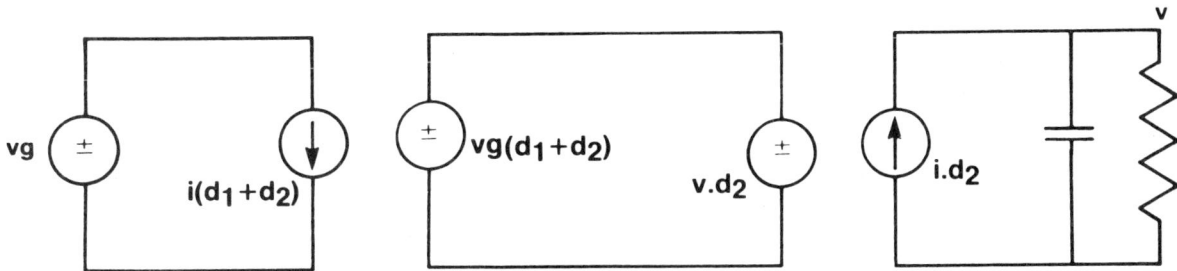

Fig. 6. Equivalent circuit for (1)-(3).

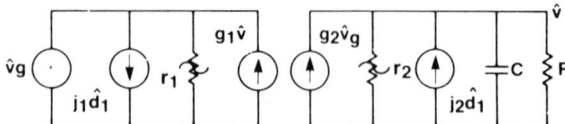

Fig. 7. Small signal low frequency linear equivalent circuit for boost converter of Fig. 3.

For the buck converter

$$\frac{\hat{V}}{\hat{V}_g} = M \frac{1}{1 + S/W_p}$$

$$\frac{\hat{V}}{\hat{d}} = \frac{2V}{M} \left(\frac{1-M}{2-M}\right)\left(\sqrt{\frac{1-M}{K}}\right) \frac{1}{1 + S/W_p}$$

where

$$M = \frac{V}{V_g} = \frac{D_1}{D_1 + D_2}$$

$$K = \frac{2L}{RT_s} \qquad W_p = \frac{2-M}{RC(1-M)}.$$

For the buckboost converter

$$\frac{\hat{V}}{\hat{V}_g} = M \frac{1}{1 + S/W_p}$$

$$\frac{\hat{V}}{\hat{d}} = \frac{V}{M\sqrt{K}} \left(\frac{1}{1 + S/W_p}\right)$$

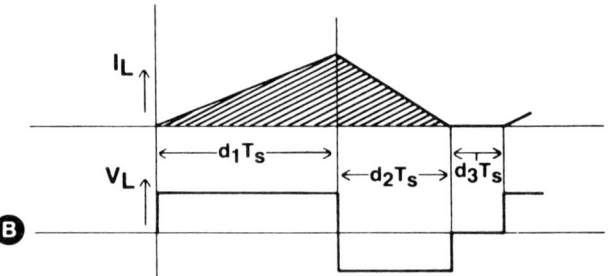

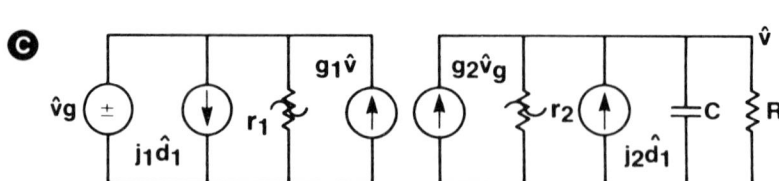

$$J_1 = \frac{2V}{R}\sqrt{\frac{1-M}{K}} \qquad r_1 = \left(\frac{1-M}{M^2}\right) R \qquad g_1 = \left(\frac{M^2}{1-M}\right) \frac{1}{R}$$

$$J_2 = \frac{2V}{RM}\sqrt{\frac{1-M}{K}} \qquad r_2 = (1-M) R \qquad g_2 = \frac{M(2-M)}{1-M} \left(\frac{1}{R}\right)$$

Fig. 8. (A) Buck converter. (B) Inductor current and voltage waveforms. (C) Its small signal low frequency linear equivalent circuit model.

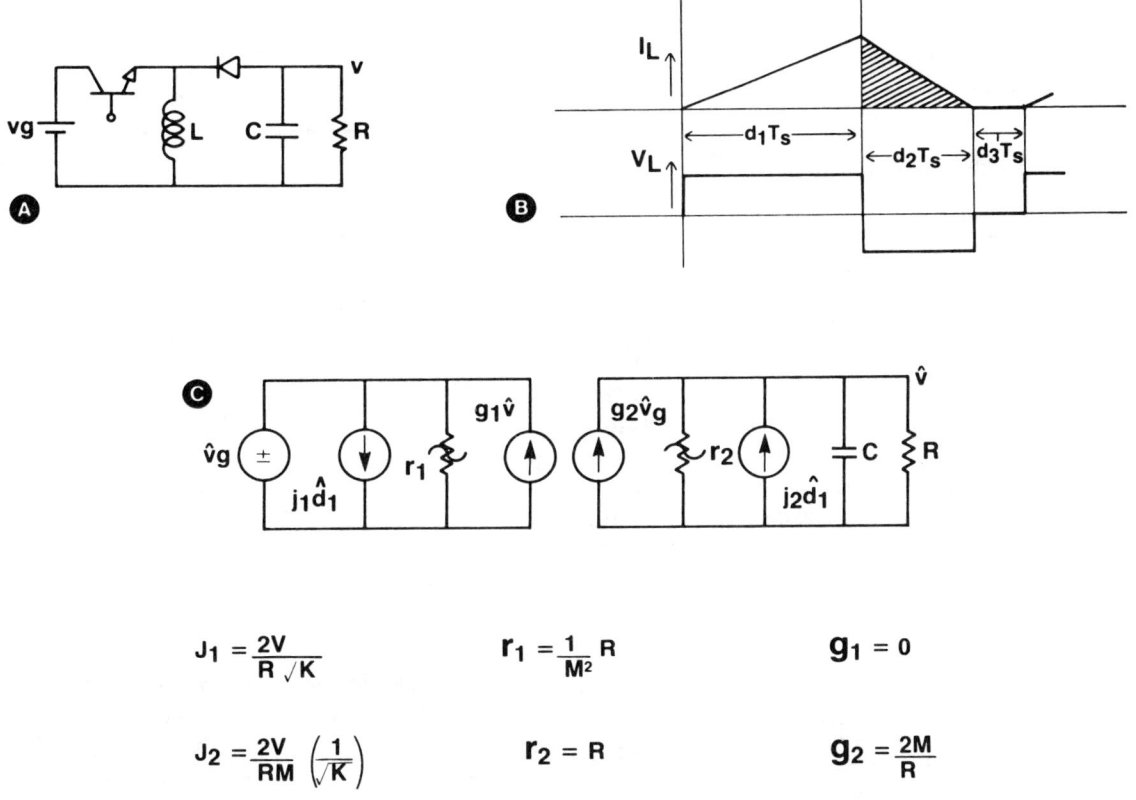

Fig. 9. (A) Buckboost converter. (B) Inductor current and voltage waveforms. (C) Its small signal low frequency linear equivalent circuit model.

where

$$M = \frac{D_1}{D_2}$$

$$K = \frac{2L}{RT_s}$$

$$W_p = \frac{2}{RC}$$

5.0 Conclusions

Having developed and demonstrated the new current injected equivalent circuit approach to modeling switching dc-dc converters operating in continuous inductor conduction mode [10], is now extended to converters operating in discontinuous inductor conduction mode. To demonstrate this approach, the modeling is carried out for buck, boost, and buckboost converters and presented in the previous sections. The results of the modeling and analysis are compared with those obtained using current injected control type approach as well as electronic equivalent circuit state space average approach.

Appendix

A. Buck Converter. Inductor current and voltage waveforms for the buck converter are shown in Fig. 8(B). Shaded portion shows the amount of current injected into the output linear circuit (parallel R and C) and the interval during which the current injected is $(d_1 + d_2)T_s$. The average inductor current injected into the output circuit during a switching period is given by

$$i_{ave} = \frac{(V_g - V)d_1 \cdot T_s}{2L}(d_1 + d_2). \quad \textbf{(1a)}$$

Volt second balance on the inductor

$$V_g(d_1) = V(d_2 + d_1). \quad \textbf{(2a)}$$

The output voltage is

$$v = (i_{ave} \cdot R)/(1 + sRC) \quad \textbf{(3a)}$$

where $R/(1 + sRC)$ is the impedance of the output network.

The steady-state conditions can now be found by using (1a)-(3a) and setting the derivative to zero; V_g, V, d_1, d_2, etc., assume steady-state values. Therefore, the above equations reduce to

$$V/V_g = \frac{D_1}{D_1 + D_2}$$

$$\frac{V}{R} = \frac{(V_g - V)D_1 T_s(D_1 - D_2)}{2L}. \quad \textbf{(4a)}$$

Equations (1a)-(3a) are perturbed around the steady-state operating point, and second order nonlinear terms are neglected once again to obtain the linear small signal model

$$\hat{i}_{ave} = \frac{T_s}{2L}\{(V_g + \hat{v}_g - V - \hat{v})(D_1 + \hat{d}_1)$$

$$\cdot (D_1 + \hat{d}_1 + D_2 + \hat{d}_2)\}$$

$$(V + \hat{v})/(V_g + v_g) = (D_1 + \hat{d}_1)/(D_1 + \hat{d}_1 + D_2 + \hat{d}_2)$$

$$\hat{V} = (R/(1 + sRC))\hat{i}_{ave}. \quad \textbf{(5a)}$$

The input to output and the control to output transfer functions are obtained from (5a) by first taking Laplace transform

$$\frac{\hat{V}}{\hat{V}_g} = (M)\frac{1}{1 + S/W_p}$$

$$\frac{\hat{V}}{\hat{d}} = \frac{2V}{M}\left(\frac{1 - M}{2 - M}\right)\sqrt{\frac{1 - M}{K}}\frac{1}{(1 + S/W_p)}$$

where

$$M = \frac{V}{V_g} = \frac{D_1}{D_1 + D_2}$$

$$K = \frac{2L}{R \cdot T_s}$$

$$W_p = \left(\frac{2 - M}{1 - M}\right)\frac{1}{RC}.$$

Using (5a), an equivalent circuit is drawn as shown in Fig. 8(C).

B. Buck-Boost Converter. Inductor current and voltage waveforms for the buckboost converter are shown in Fig. 9(B). Shaded portion shows the amount of current injected into the output linear circuit (parallel R and C), and the interval during which the current injected is $d_2 T_s$. The average inductor current injected into the output circuit during a switching period is given by

$$i_{ave} = \frac{V_g \cdot d_1 \cdot T_s \cdot d_2}{2L} \quad \textbf{(1b)}$$

Volt second balance on the inductor

$$V_g(d1) = V \cdot d_2. \quad \textbf{(2b)}$$

The output voltage is

$$v = (i_{ave} \cdot R)/(1 + sRC) \quad \textbf{(3b)}$$

where $R/(1 + sRC)$ is the impedance of the output network. The steady-state conditions can now be found by using (1b)-(3b) and setting the derivative to zero; V_g, V, d_1, d_2, etc., assume steady-state values. Therefore, the above equations reduce to

$$V/V_g = \frac{D_1}{D_2} \quad \frac{V}{R} = \frac{V_g \cdot D_1 \cdot D_2 \cdot T_s}{2L} \quad \textbf{(4b)}$$

Equations (1b)-(3b) are perturbed around the steady-state operating point and second order nonlinear terms are neglected once again to obtain the linear small signal model

$$\hat{i}_{ave} = \frac{T_s}{2_L}(\hat{V}_g \cdot D_1 \cdot D_2 + V_g \cdot \hat{d}_1 \cdot D_2 + V_g \cdot D_1 \cdot \hat{d}_2)$$

$$\hat{V}/V_g - V \cdot \hat{V}_g/V_g^2 = \hat{d}_1/D_2 - D_1 \cdot \hat{d}_2/D_2^2$$

$$\hat{V} = (R/(1 + sRC))\hat{i}_{ave.} \quad \textbf{(5b)}$$

The input to output and the control to output transfer functions are obtained from (5b) by first taking Laplace transform

$$\frac{\hat{V}}{\hat{V}_g} = (M) \frac{1}{1 + S/W_p}$$

$$\frac{\hat{V}}{\hat{d}} = \frac{V}{M\sqrt{K}} \frac{1}{1 + S/W_p}$$

where

$$M = \frac{V}{V_g} = \frac{D_1}{D_2}$$

$$K = \frac{2L}{R \cdot T_s}$$

$$W_p = \frac{2}{RC}$$

Using (5b), an equivalent circuit is drawn as shown in Fig. 9(C).

References

1. R.D. Middlebrook and S. Ćuk, "Modeling and analy. methods for dc-dc switching converters," presented at the IEEE Int. Semi-Conductor Power Converter Conf., (Orlando, FL), 1977.
2. A.J. Fossard and M. Clique, "Modelisation des cellules elementaires," Tech. Rep. 1, ESTEC contract 2590/75 AK, 1976.
3. R.D. Middlebrook and S. Ćuk, "A general unified approach to modeling switching converter power stages," in *IEEE Power Electron. Specialists Conf., 1976 Rec., pp. 18-34.*
4. S. Ćuk and R.D. Middlebrook, "A general unified approach to modeling switching dc-to-dc converters in discontinuous conduction mode," presented at the IEEE Power Electronics Specialists Conf., June 1977.
5. S.-P. Hsu et al., "Modeling and analysis of switching dc-dc converters in constant frequency current programmed mode," in *IEEE Power Electron. Specialists Conf. 1979 Rec.*
6. Art Brown, "State Space Analysis of pulse width modulated switching converters," Power Electronics Group, California Institute of Technology, technical note T-58, Feb. 6, 1979.
7. A.J. Fossard and M. Clique, "Modeling and design of dc-dc converters using modern control theory, Part 1 Modelization, Part 2 Open Loop Analysis and Control Design," in *Proc. Third ESTEC Spacecraft Power Conditioning Seminar,* (Noordwijk. The Netherlands), 1977.
8. A.J. Fossard et al., "A general linear continuous model for design of power conditioning units at fixed and free running frequency," in *IEEE Power Electronics Specialists Conf., 1977 Rec.,* pp. 113-124.
9. R. Prajoux, J.C. Marpinard, and J. Jalade, "Establishment de modeles mathematiques pour regulaterus de puissance a modulation de largeur d impulsions (pwm); 2. *Models continus,*" ESA Sci. Tech. Rev., vol 2, no. 1, 1966, pp. 115-129.
10. P.R.K. Chetty, "Current injected equivalent circuit approach (CIECA) to modeling of switching dc-dc converters," *IEEE Trans. Aerospace and Electronic Syst.,* vol. 17, Nov. 1981.

CIECA: APPLICATION TO CURRENT PROGRAMMED SWITCHING DC-DC CONVERTERS

The current injection equivalent circuit approach (CIECA) to modeling switching converter power stages is extended to model the current programmed converter power stages operating in fixed frequency, continuous inductor conduction mode. To demonstrate the method, modeling is carried out for the buck, boost, and buckboost converters to obtain small-signal linear equivalent circuit models which represent both input and output properties. The results of these analyses are presented in the form of linear equivalent circuit models as well as transfer functions. Though current programmed converters exhibit single-pole response, the addition of artificial ramp changes converters to exhibit well damped two-pole response. This has been investigated for the first time using CIECA. The results of these analyses are presented in the form of linear equivalent circuit models as well as transfer functions.

1.0 Introduction

In the last ten years, modeling of switching dc-dc converters has received considerable attention and the effort has resulted in the characterization of transfer as well as input and output properties of basically nonlinear switching dc-dc converters in the frequency domain. Among the various approaches attempted to attain this goal, the current injected equivalent circuit approach (CIECA) is very versatile [1,2]. This approach is now extended to current programmed converters operating in the continuous inductor conduction (CIC) mode. The advantages of the current programming are already well known and are summarized as follows:

1) Switching converter active components are protected from excessive overload and stress. This allows controlled derating of components.
2) Switching converters when current programmed behave basically as first-order systems.
3) Several converters can be operated in parallel without load sharing problems.
4) Inductor sawtooth current waveform replaced advantageously the reference sawtooth necessary to generate the pulsewidth-modulated control signal.

Thus this paper is concerned with current programmed switching dc-dc converters operating in the fixed frequency, CIC mode. The main goal is to obtain the small-signal equivalent circuit models which represent both input and output properties, which can then be embedded in the model of a complete regulator system, so that the overall dynamic properties and the stability can be analyzed and designed.

CIECA is briefly reviewed in Section 2. Section 3 contains the explanation for the instability of the current programmed converters when they operate at duty ratios greater than 0.5 Section 4 contains the detailed development of modeling for boost converters. Following the same approach, modeling is carried out for buck and buckboost converters and the results are presented in Section 5. Section 6 presents modeling for the current programmed stabilized converters. Section 7 compares the results with those obtained using other modeling approaches [3,4]. The salient feature of the current programmed CIC mode is that the control-to-output transfer function is basically a one-pole function if the effect of adding an artificial ramp to stabilize the converter is neglected (see Section 8). One more interesting investigation made for the first time is that the effect of the inclusion of an artificial ramp adds back the pole but with large damping. Section 9 presents the conclusions on the results of the work carried out.

2.0 Review of CIECA

The current injected equivalent circuit approach to modeling switching converters in duty ratio programmed mode has been developed [1,2]. Since this approach [1] is used in this paper for modeling and analysis of converters operated in the

---

© 1982 IEEE. Reprinted with permission from *IEEE Transactions on Aerospace and Electronic Systems*, Vol. AES-18, No. 5, pp. 538-544, Sept. 1982.

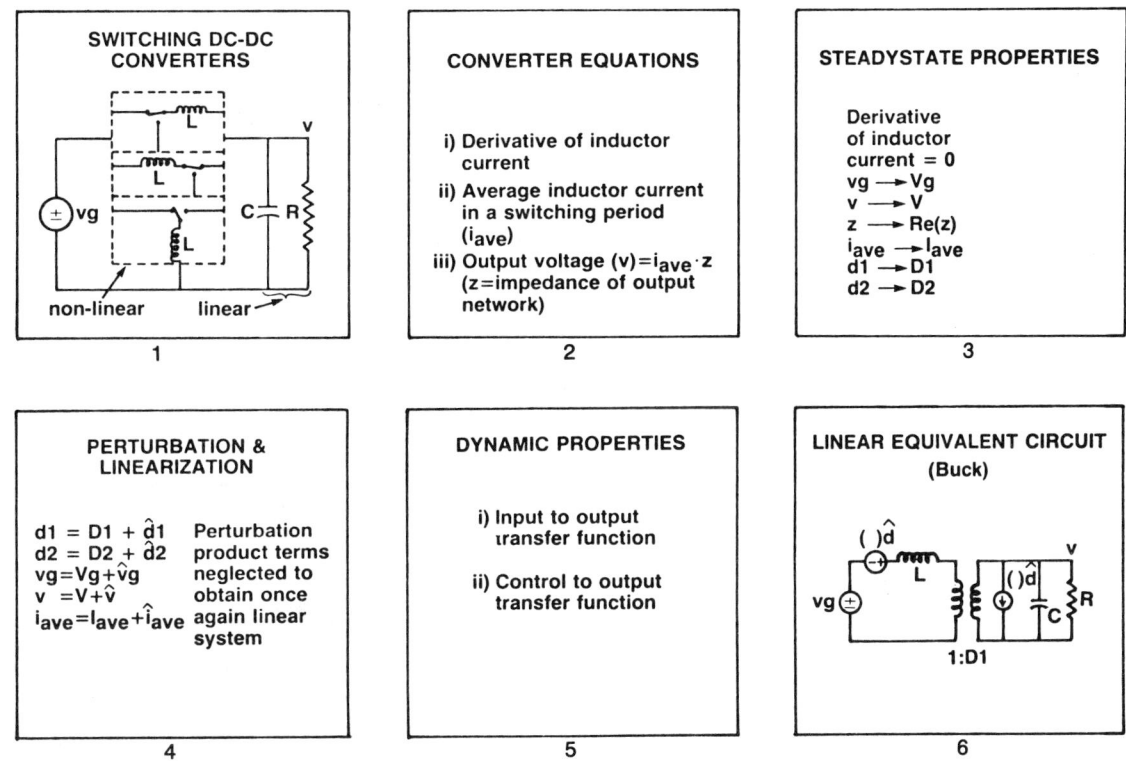

Fig. 1. Flowchart of modeling switching dc-to-dc converters in the CIC mode using CIECA.

fixed frequency mode, a brief review of this approach is presented here. Of course the same approach can be also used in modeling the converters operated in variable frequency mode.

The following conventions and notations are followed in the modeling and analysis: $d1Ts$ is the interval during which the transistor is turned-on and the diode is off; $d2Ts$ is the interval during which the transistor is turned-off and the diode is on; $d1Ts + d2Ts = T_s$, and $Ts = 1/f_s$ is the switching period. Capitalized quantities indicate steady-state values and quantities with carets indicate small perturbations.

Modeling converters operating in the CIC mode using CIECA is outlined in the flowchart of Fig. 1, which is very general and is applicable to various power stages. The first step in this process is to identify the nonlinear and linear parts of the converter circuit and to linearize only the nonlinear part of the converter as the remainder of the converter is inherently linear (box 1). The nonlinear part of the converter determines the average current injected into the linear part. Now (box 2) a set of relationships are written referring to the converter diagram and the current and voltage waveforms shown in Fig. 2:

1) average current (i_{ave}) injected into the linear part in a switching period;
2) derivative of the inductor current function of the value of the inductor, the voltage across that in a switching period;
3) relationship between average injected current and output voltage $v = (i_{ave}) \times (z)$ where z is the impedance of the linear part of the converter.

Now steady-state solution is achieved by setting derivatives and perturbations to zero (box 3). Since the converter equations in box 2 are linear,

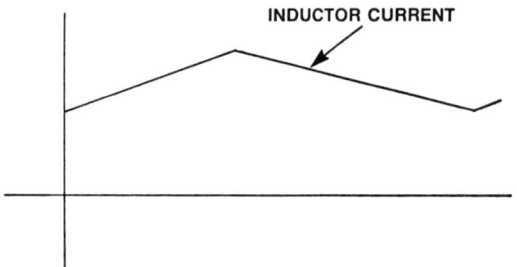

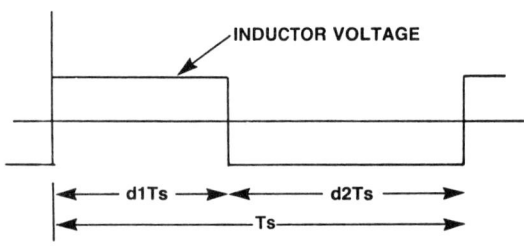

Fig. 2. Typical inductor current and voltage waveforms in buck converter.

superposition holds and can be perturbed (box 4) by the introduction of a small ac variation over the steady-state operating point. As we know, the independent driving inputs are v_g and d, the perturbation in these two inputs cause the perturbation in i and v. Now by making the small-signal approximation, namely, the small ac variation from the steady-state operating point being negligible compared with the steady-state operating point values, $\hat{v}/V$, $\hat{v}_g/V_g$, $\hat{d}_1/D_1$, $\hat{d}_2/D_2$, $\hat{i}/I$, (each) $\ll 1$. Using these approximations, nonlinear second-order terms are neglected to obtain once again a linear set of equations. Now only the ac part is retained which describes the small-signal low-frequency behavior of the converter. Using this set of equations, the input-to-output and control-to-output transfer functions (box 5) are written. Using the same set of equations, an equivalent circuit (box 6) is drawn which represents the input and output small-signal low-frequency properties of the nonlinear converter.

3.0 Instability in Current Programmed Converters

As mentioned in Section 1, current programmed converters operating in fixed frequency CIC mode exhibit instability even in the absence of external feedback to regulate the output [4]. This instability occurs when the duty ratio exceeds 0.5 and the cause for this instability is that the current programming itself constitutes an internal feedback, the gain of which becomes positive and attains a value of 1 at a duty ratio of 0.5 and increases as the duty ratio increases. Thus the current programmed converters, even in the open loop, exhibit

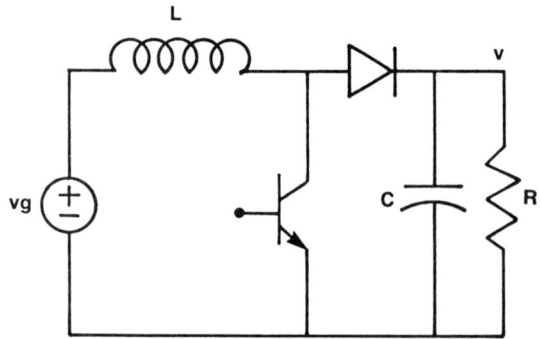

Fig. 3. Boost converter with all parasitics and storage time effects neglected.

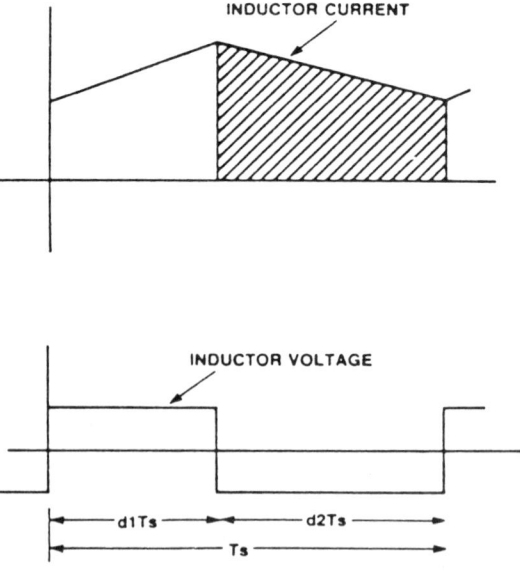

Fig. 4. Inductor current and voltage waveforms of the boost converter.

instability when the duty ratio exceeds 0.5.

4.0 Modeling of Boost Converter

Modeling converters using CIECA, as reviewed in Section 2, is applied to the boost converter and current programming is introduced in the process. With the assumption of ideal switches (no parasitics or storage time modulation effects have been considered), the converter diagram is shown in Fig. 3. Inductor current and voltage waveforms for the boost converter are shown in Fig. 4. The shaded portion shows the amount of current injected into the output linear circuit (parallel R and C) and the interval during which the current injected is $d2Ts$. The average inductor current injected into the output circuit during a switching period is given by

$$i_{ave} = d2i \quad (1)$$

where i is the average inductor current. The derivative of the inductor current is given by

$$L(di/dt) = Vg - d2 \cdot V. \quad (2)$$

The output voltage is

$$V = i_{ave} \cdot [R/(1 + SRC)] \quad (3)$$

where $R/(1 + SRC)$ is the impedance of the output network. In the boost converter, the programmed current is actually the inductor current since that is the current which flows through the switch when it is turned on. Therefore i is constrained with the control signal i_c,

$$i_c = i \quad (4)$$

The steady-state conditions can now be found by using (1)-(3), setting frequency terms to zero, and setting all other quantities to their steady-state values. Therefore the above equations reduce to

$$V/Vg = 1/D2 = M \quad (5)$$

$$I = V/(R \cdot D2).$$

Equations (1)-(4) are perturbed around the steady-state operating point, and second-order nonlinear terms are neglected once again to obtain the linear small-signal model.

$$\hat{i}_c = \hat{i} \quad (6)$$

$$\hat{i}_{ave} = D2 \cdot \hat{i} - I \cdot \hat{d} \quad (7)$$

$$L(d\hat{i}/dt) = \hat{V}g - D2 \cdot \hat{V} + V \cdot \hat{d} \quad (8)$$

$$\hat{V} = [R/(1 + SRC)]\hat{i}_{ave}. \quad (9)$$

After taking the Laplace transform of (8), $\hat{d}$ can be written as

$$\hat{d}(s) = sL\hat{i}(s)/V + D2 \cdot \hat{V}(s)/V - \hat{V}g(s)/V. \quad (10)$$

The line-to-output and control-to-output transfer functions can be written from the above equations as

$$\hat{V}/\hat{V}g = (1/2D2)[1/(1 + SRC/2)]$$

$$\hat{V}/\hat{i}_c = (R \cdot D2/2)$$
$$\cdot (1 - SL/(R\ D2^2)/(1 + SRC/2). \quad (11)$$

The salient feature of the result is now apparent. Both responses show single pole response when the effect of adding an artificial ramp is excluded. An equivalent linear circuit model is developed using (7)-(10) as shown in Fig. 5 which completely describes the input and output properties of the cur-

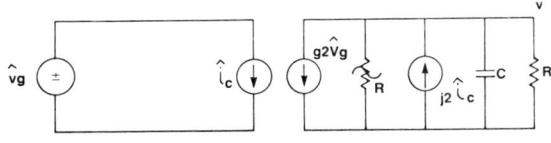

Fig. 5. Linear equivalent circuit model for the current programmed boost converter in CIC mode.

rent programmed nonlinear boost converter operating in the CIC mode.

Section 6 presents the modeling of current programmed stabilized converters, i.e., to include the effect of adding an artificial ramp.

5.0 Buck and Buckboost

The same method has been followed to model the current programmed buck and buckboost converters operating in the CIC mode. The converter and its equivalent circuit are shown in Fig. 6 for the buck converter and in Fig. 7 for the buckboost converter. In both cases the switch current is the inductor current during the interval $d1 Ts$. The equivalent circuits of Fig. 6(B) and 7(B) contain familiar current sources driving the RC network, i.e., single-pole response in both cases: the current modulation generators as a function of i_c, the control signal, and vg, the input modulation voltage and the filter C in parallel with load R.

The result for the two transfer functions of major interest, the line-to-output transfer function and control-to-output transfer function, are as follows.

For the buck converter, $\hat{v}$ does not depend upon Vg,

$$\hat{V}/\hat{i}_c = R/(1 + SRC). \quad (12)$$

For the buckboost converter,

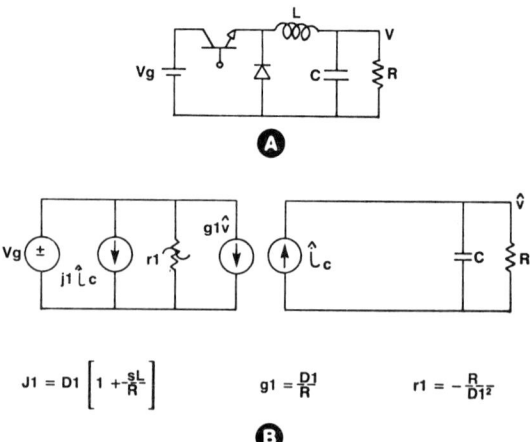

Fig. 6. (A) Buck converter. (B) Small signal low frequency linear equivalent circuit model of buck converter.

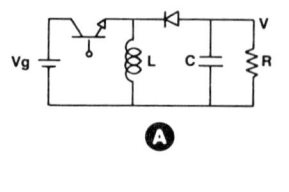

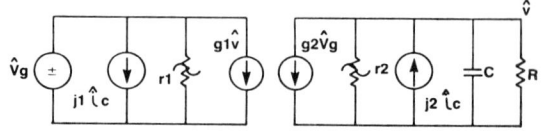

Fig. 7. (A) Buckboost converter. (B) Low frequency small signal linear equivalent circuit model of buckboost converter.

$$\hat{V}/\hat{V}_g = (D1/D2)^2 [1/(1 + SRC/D2)]$$

$$\hat{V}/\hat{i}_c = R(1 - SLeD/R)/(1 + SRC/D2)$$

$$Le = L/(D2)^2. \quad (13)$$

As in the case of the boost converter, the buck and buckboost converters exhibit one-pole response and it is interesting to note that inclusion of an artificial ramp into the modeling changes all these results, which is dealt with in Section 6.

6.0 Modeling of Stabilized Current Programmed Boost Converter

As mentioned in Section 3, an artificial ramp with suitable slope has been added to the switching current to stabilize the converter even in the absence of external feedback. Figure 8 shows the

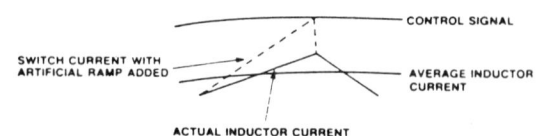

Fig. 8. Actual inductor current in boost converter, control signal and artificial ramp of appropriate slope m added to switch current.

actual inductor current, the control signal, and an artificial ramp of appropriate slope m added to the switch current. Now looking at the waveform, the average inductor current i can be related to i_c as

$$i = i_c - m \cdot d1 \cdot Ts - m1 \cdot d1 \cdot Ts/2. \quad (14)$$

Perturbing (14) and retaining only ac terms

$$\hat{i} = \hat{i}_c - (m + m1/2) Ts \cdot \hat{d} - (D1 \cdot Ts/2)\hat{m}1 \quad (15)$$

where $m1$ is the slope of the inductor current and m is the slope of the artificial ramp added to stabilize the open loop converter.

For the boost converter $m1$ is given by

$$m1 = Vg/L \quad (16)$$

where m is chosen to be the negative slope of the programmed (inductor) current and is given by

$$m = (V - Vg)/L. \quad (17)$$

Using (16) and (17), (15) is rewritten after taking Laplace transform as

$$\hat{i}(s) = \hat{i}_c(s) - [V(2 - D2)/RK]\,\hat{d}(s)$$

$$- (D/RK)\,\hat{V}_g(s) \quad (18)$$

Substituting the value of $\hat{i}$ from (18) into (10) yields

$$\hat{d}(s) = \{ (SL/V)\hat{i}_c(s) + (D2/V)\hat{V}(s)$$

$$- (1/L)(1 + SLD/RK)\,\hat{V}g(s)\}/$$

$$[1 + SL(2 - D2)/RK]. \quad (19)$$

Line-to-output and control-to-output transfer functions can be written using (7), (9), and (19) as below:

$$\hat{V}(s)/\hat{V}g(s) = Ag\,[(1 + S/Wz1)/(1 + S/WQ + S^2/W^2)]$$

$$\hat{V}(s)/\hat{i}_c(S) = Ac\,[(1 - S/Wz2)/(1 + S/WQ + S^2/W^2)].$$

Under the assumption that $D1, D2 \ll K$ (converter being operating deep in the CIC mode),

$$W_{z1} = 2/D1 \cdot Ts$$

$$Ag = 1/2 \cdot D2$$

$$W_{z2} = R(D2)^2/L$$

$$Ac = R \cdot D2/2$$

$$W^2 = 2K/L \cdot C(2 - D^2).$$

From the above equations, it is clear that either the control-to-output or the line-to-output transfer functions exhibit two-pole response. This is quite in contrast to the single-pole response of current programmed converters in the absence of an artificial ramp.

7.0 Physical Explanation

The following explanation applies to the current programmed converters operating in the CIC mode and in the absence of an artificial ramp with appropriate slope. Of course the converter will be unstable if the duty ratio is greater than 0.5. When the converter is current programmed, the state variable loses its contribution toward a pole due to inductor current. This happens because the inductor current is no longer an independent variable and is constrained by the control signal. Though the development of inductor current depends upon the value of the inductor and other operating parameters, its magnitude is constrained by the control signal.

Now consider the current programmed converter operating in the CIC mode, but in the presence of an artificial ramp. Of course, now the converter will be stable throughout the operating range of duty ratios. Refer to the schematic of Fig. 9 where the practical implementation of current programming and addition of an artificial ramp are shown. The switch $S1$ in Fig. 9 is purposely included for better explanation, Rs is the resistor

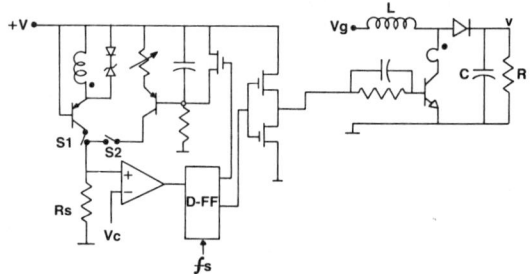

Fig. 9. Practical implementation of current programmed boost and addition of artificial ramp.

across which a current proportional to the switch current is produced when S1 is closed and closure of S2 adds an artificial ramp to this current, the slope of which can be adjustable or continuously programmable in a high performance system. When S2 is open and S1 is closed, the converter will be stable only if the duty ratio is less than 0.5. To make the converter stable over the entire range of duty ratio, an artificial ramp is added by closing the switch S2.

Now consider the same situation in two steps.

Step 1: Only current programming is present, i.e., S1 is closed and S2 is open.

Step 2: Only an artificial ramp is present, i.e., S1 is open and S2 is closed.

We discussed Step 1 at the beginning of this section. The current programmed converters in the absence of artificial ramp exhibit single-pole control-to-output response.

Now consider Step 2. In the absence of current proportional to switch current, only the artificial ramp is present. This means the duty ratio is constant in the absence of switch current and hence the converter behaves as if it is duty ratio programmed. Everybody knows that duty ratio programmed converters (buck, boost, buckboost) exhibit two-pole response.

To get the overall effect, Steps 1 and 2 have to be summed up in proper perspective. The net effect is that the converters now exhibit two-pole response but well (over) damped. This also can be seen as the two poles being well separated as shown in Fig. 10. Thus, though current programmed converters exhibit single-pole response in the absence of the artificial ramp, they exhibit two-pole response in the presence of the stabilizing artificial ramp. This has been investigated for the first time using CIECA. This effect is very significant as sophisticated and high performance is expected from the power processing systems.

8.0 Modeling of Stabilized Current Programmed Buck and Buckboost Converters

The same method has been followed to model the stabilized current programmed buck and buckboost converters operating in the CIC mode. The results for the two transfer functions of major interest, the line-to-output transfer function and control-to-output transfer function, are as follows.

For the buck converter,

$$\hat{V}(s)/\hat{V}g(s) = Ag[1/(1 + S/WQ + S^2/W^2)]$$

$$\hat{V}(s)/\hat{i}_c(s) = Ac[1/(1 + S/WQ + S^2/W^2)].$$

Under the assumption $D1, D2 \ll K$,

$$Ag = -D/K$$

$$Ac = R$$

$$W^2 = K/LC.$$

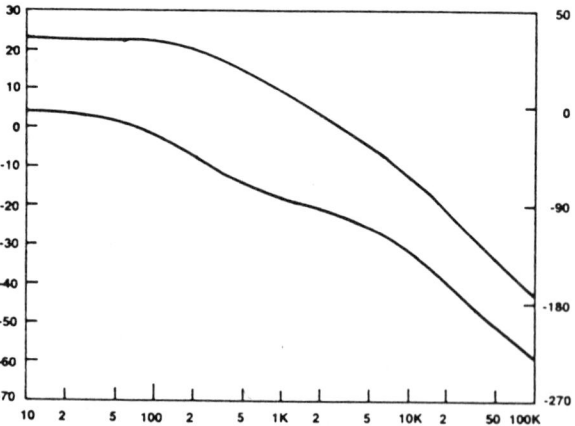

Fig. 10. Bode plot of control to output transfer function of boost converter.

For the buckboost converter,

$$\hat{V}(s)/\hat{V}_g(s) = A_g[(1 + S/W_{z1})/(1 + S/WQ + S^2/W^2)]$$

$$\hat{V}(s)/\hat{i}_c(s) = A_c[(1 + S/W_{z2})/(1 + S/WQ + S^2/W^2)].$$

Under the assumption that $D1, D2 \ll K$,

$$W^2 = K/LC(1 - 2D/D2).$$

As in the case of the boost converter, the buck and buckboost converters exhibit two-pole response.

9.0 Conclusions

Modeling switching converters using CIECA has been extended to current programmed converters operating in the CIC mode. To demonstrate the approach, modeling is carried out for buck, boost, and buckboost converters. Though the converters exhibit single-pole response in the absence of a stabilizing artificial ramp, they exhibit well (over) damped two-pole response when the effect of an artificial ramp is included. For the first time such an effect has been investigated. This investigation is highly significant and is very important for high performance power processing systems.

Thus the modeling developed for converters in fixed frequency current programmed CIC mode using CIECA permits us to design regulators containing these converter power stages to achieve required performance.

References

1. Chetty, P.R.K. Current injected equivalent circuit approach (CIECA) to modeling of switching dc-dc converters in continuous inductor conduction mode. Internal Report, Sundstrand Advanced Technology Corporation, Rockford, IL.
2. Chetty, P.R.K. Current injected equivalent circuit approach (CIECA) to modeling of switching dc-dc converters in discontinuous inductor conduction mode. Internal Report, Sundstrand Advanced Technology Corporation, Rockford, IL.
3. Hsu, S.-P., Brown, A., Rensink, L., and Middlebrook, R.D. (1979). Modeling and analysis of switching dc-to-dc converters in constant frequency current programmed mode. *IEEE Power Electronics Specialists Conference Record*, 1979.
4. Capel, A., Clique, M., and Fossard, A.J. (1980). Current control modulators: general theory on specific designs. *IEEE Power Electronic Specialists Conference Record*, 1980.

CURRENT INJECTED EQUIVALENT CIRCUIT APPROACH TO MODELING AND ANALYSIS OF CURRENT PROGRAMMED SWITCHING DC-DC CONVERTERS (DISCONTINUOUS INDUCTOR CONDUCTION MODE)

The current injected equivalent circuit approach (CIECA) to modeling switching converter power stages is extended to model the current programmed converter power stages operating in fixed frequency discontinuous inductor conduction mode. To demonstrate the method, the modeling is carried out for the buck, boost, and buckboost converters to obtain small signal linear equivalent circuit models that represent both input and output properties. The results of these analyses are presented in the form of linear equivalent circuit models as well as transfer functions.

1.0 Introduction

In the last ten years, modeling of switching dc-dc converters has received considerable attention, and the effort has resulted in characterization of transfer as well as input and output properties of basically nonlinear switching dc-dc converters in the frequency domain. Among the various approaches attempted to attain this goal, the current injected equivalent circuit approach (CIECA) has the merits of both electronic equivalent circuit state space average approach and current injected control theory approach [5], [6], and its devoid of demerits.

© 1982 IEEE. Reprinted with permission from *IEEE Transactions on Industry Applications*, Vol.IA-18, No.3, pp. 295-299, May/June 1982.

This approach, having extended to current programmed converters operating in continuous inductor conduction (CIC) mode [3], is now applied to model the current programmed switching converters operating in discontinuous inductor conduction (DIC) mode. The advantages of current programming are already well known and are summarized below.

1) Switching converter active components are protected from excessive overload and stress. This allows controlled derating of components.
2) Switching converters when current programmed behave basically as first-order systems.
3) Several converters can be operated in parallel without load sharing problems.
4) Inductor sawtooth current waveform replaces advantageously the reference sawtooth necessary to generate the pulsewidth modulated control signal.

Thus, this paper is concerned with current programmed switching dc-dc converters operating in the fixed frequency discontinuous inductor conduction mode. The main goal, as in the previous work [1-3], is to obtain the small signal equivalent circuit models that represent both input and output properties, which can then be embedded in the model of a complete regulator system so that the overall dynamic properties and the stability can be analyzed and designed.

The current injected equivalent circuit approach [1-3] followed in this analysis and modeling is briefly reviewed in Section 2. Section 3 contains the detailed development of modeling for boost converter. Following the same approach, the modeling is carried out for buck and buckboost converters and the results are presented in Section 4. The results are compared with those obtained using state space average electronic equivalent circuit approach and current injected control theory approach. The salient feature of the current programmed DIC mode is that the control-to-output transfer function is basically a one pole, as in the current programmed CIC mode. One more interesting investigation that made, as in [4], is that the buck converter in this current programmed DIC mode goes into oscillations under certain steady-state operating conditions.

The cause for this is discussed, and a remedy is suggested and implemented successfully to eliminate this potential instability in [4]. The final section presents conclusions on results of the work carried out.

2.0 Review

The current injected equivalent circuit approach to modeling switching converters in duty ratio programmed mode has been developed [1], [2]. Since this approach [2] is used in this paper for modeling and analysis of converters operated in the fixed frequency mode, a brief review is presented here. Of course, the same approach can be also used in modeling the converters operated in variable frequency mode.

The following conventions and notations are followed in the modeling and analysis.

$d1Ts$ the interval during which the transistor is turned on and the diode is off
$d2Ts$ the interval during which the transistor is turned off and the diode is on
$d3Ts$ the interval during which the transistor is turned off and the diode is off.

$$d1Ts + d2Ts + D3Ts = Ts$$

$$\text{and } Ts = 1/f_s$$

$$= \text{switching period.}$$

The capitalized quantities are used for steady-state values and the quantities with hats for the small perturbations.

The current injected equivalent circuit approach to modeling converters operating in the discontinuous inductor conduction mode is outlined in the flowchart of Fig. 1, which is very general and applicable to various power stages. The first step in this process is to identify the nonlinear and linear parts of the converter circuit and linearize only

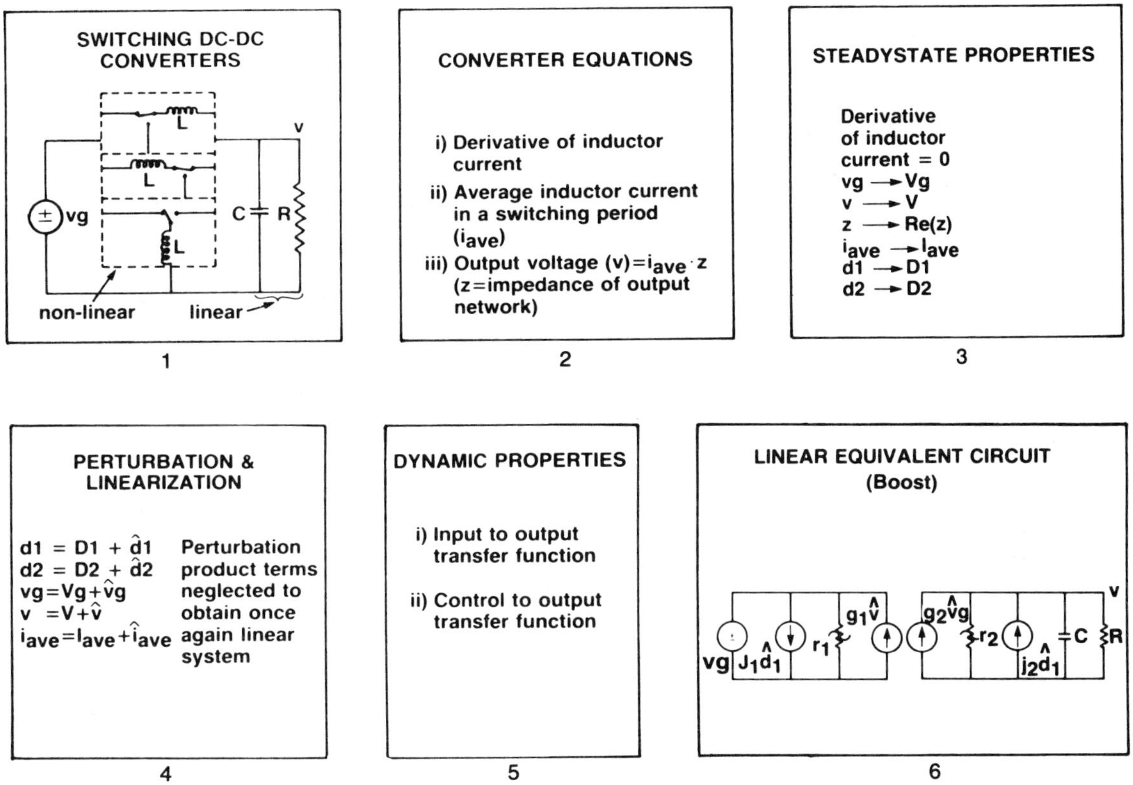

Box 1: SWITCHING DC-DC CONVERTERS
(non-linear | linear)

Box 2: CONVERTER EQUATIONS
i) Derivative of inductor current
ii) Average inductor current in a switching period (i_{ave})
iii) Output voltage $(v) = i_{ave} \cdot z$ (z = impedance of output network)

Box 3: STEADYSTATE PROPERTIES
Derivative of inductor current = 0
$v_g \rightarrow V_g$
$v \rightarrow V$
$z \rightarrow Re(z)$
$i_{ave} \rightarrow I_{ave}$
$d1 \rightarrow D1$
$d2 \rightarrow D2$

Box 4: PERTURBATION & LINEARIZATION
$d1 = D1 + \hat{d}1$
$d2 = D2 + \hat{d}2$
$v_g = V_g + \hat{v}_g$
$v = V + \hat{v}$
$i_{ave} = I_{ave} + \hat{i}_{ave}$

Perturbation product terms neglected to obtain once again linear system

Box 5: DYNAMIC PROPERTIES
i) Input to output transfer function
ii) Control to output transfer function

Box 6: LINEAR EQUIVALENT CIRCUIT (Boost)

Fig. 1. Flowchart of current injected equivalent circuit approach to modeling switching dc-dc converters in DIC mode.

the nonlinear part of the converter, as the remaining of the converter is inherently linear (box 1). The nonlinear part of the converter determines the average current injected into the linear part. Now (box 2) a set of relationships is written referring to the converter diagram and current and voltage waveforms shown in Fig. 2.

1) Volt-second balance on the inductor.
2) Average current (i_{ave}) injected into the linear part in a switching period.
3) Relationship between average injected current and output voltage $v = (i_{ave}) \times (z)$, where z is the impedance of the linear part of the converter.

Now the steady-state solution is achieved by setting derivatives and perturbations to zero (box 3). Since the converter equations in box 2 are linear, superposition holds and can be perturbed (box 4) by the introduction of a small ac variation over the steady-state operating point. As we know the independent driving inputs are v_g and d, so the perturbation in these two inputs cause the perturbation in i and v. Now making the small signal approximation, namely, the small ac variation from the steady-

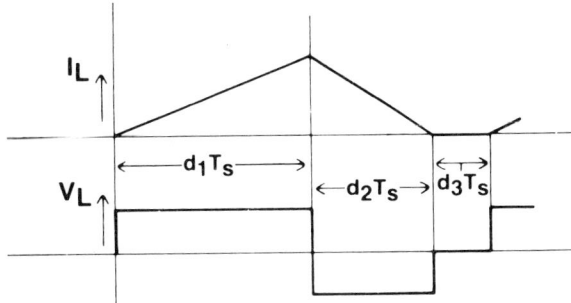

Fig. 2. Typical inductor current and voltage waveforms in buck converter.

state operating point are negligible compared to the steady-state operating point values, i.e.,

$\hat{v}/V, \hat{v}g/Vg, \hat{d}1/D1, \hat{d}2/D2, \hat{i}/I$ (each) $\ll 1$.

Using the above approximations, nonlinear second-order terms are neglected to obtain once again a linear set of equations. Now only the ac part is retained which describes the small signal low frequency behavior of the converter. Using this set of equations, the input-to-output and control-to-output transfer functions (box 5) are written. Using the same set of equations an equivalent circuit (box 6) is drawn which represents the input and output small signal low frequency properties of the nonlinear converter.

3.0 Modeling of Boost Converter

The current injected equivalent circuit approach to modeling converters reviewed in the previous section is applied to boost converter, current programming being introduced in the process. With the assumption of ideal switches (no parasites or storage time modulation effects have been considered), the converter diagram is shown in Fig. 3.

Inductor current and voltage waveforms for the boost converter are shown in Fig. 4. The shaded portion shows the amount of current injected into the output linear circuit (parallel R and C) and the interval during which the current injected is $d2Ts$. The average inductor current injected into the out-

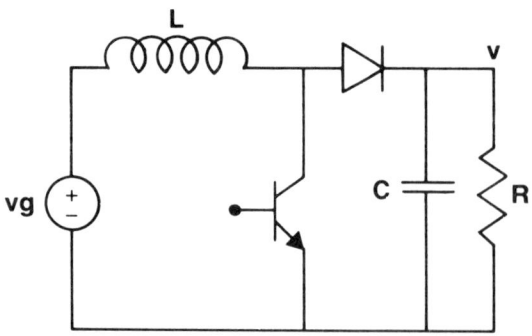

Fig. 3. Boost converter with all parasitics and storage time effects neglected.

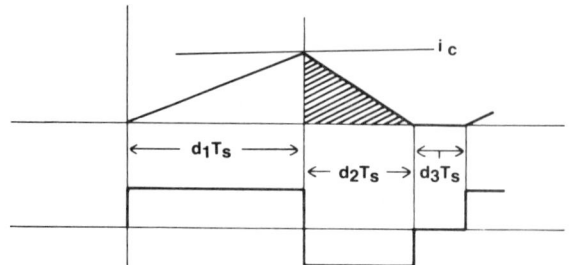

Fig. 4. Inductor current and voltage waveforms of the boost converter.

put circuit during a switching period is given by

$$i_{ave} = d2i$$

where i is the average inductor current.

$$i = \frac{Vgd1Ts}{2L}, \quad (1)$$

Volt-second balance on the inductor is

$$Vg(d1 + d2) = V \cdot d2. \quad (2)$$

The output voltage is

$$V = i_{ave} \cdot \{R/(1 + SRC)\} \quad (3)$$

where $R/(1 + SRC)$ is the impedance of the output network. In the boost converter, the programmed current is actually the inductor current since that is the current which flows through the switch when it is turned on. Therefore, constrain i with the control signal i_c as i_c constrains i peak,

$$i_c = i \text{ peak} = \frac{Vg \cdot d1Ts}{L}.$$

This can be rewritten in terms of $d1$:

$$d1 = \frac{L \cdot i_c}{Vg \cdot Ts}. \quad (4)$$

The steady-state conditions can now be found by using (1)-(4) and setting frequency terms to zero and all other quantities to their steady-state values.

Therefore the above equations reduce to

$$\frac{V}{V_g} = 1 + \frac{D1}{D2} = M$$

$$I_{ave} = \frac{V_g \cdot D1 T_S \cdot D2}{2L} \quad (5)$$

$$V = I_{ave} \cdot R.$$

Equations (1)-(4) are perturbed around the steady-state operating point and second-order nonlinear terms are neglected once again to obtain the linear small signal model.

$$\frac{\hat{V}}{V_g} - \frac{V}{V_g^2}\hat{V}_g = \frac{1}{D2}\hat{d}_1 - \frac{D1}{D_2^2}\hat{d}_2 \quad (6)$$

$$\hat{V} = \frac{T_s}{2L}\{D1 D2 \hat{V}_g + D2 V_g \hat{d}_1 + V_g D1 \hat{d}_2\} \quad (7)$$

$$\hat{d}_1 = \frac{\frac{R}{1+SRC}}{V_g \cdot T_s} L\hat{i}_c - \frac{D1}{V_g}\hat{V}_g. \quad (8)$$

Eliminating $\hat{d}_2$ using (6), $\hat{d}_1$ using (8) and (7), and taking the Laplace transform yields the following simplified equation:

$$\hat{V}(s)\left\{1 + \frac{SRC(M-1)}{(2M-1)}\right\}$$

$$= \frac{2R(M-1)D2}{(2M-1)}\hat{i}_c(s) + \frac{M}{2M-1}\hat{V}_g(s).$$

The line-to-output and control-to-output transfer functions can be written from the above equation as

$$\frac{\hat{V}}{\hat{V}_g} = \left(\frac{M}{2M-1}\right)\frac{1}{1+s/\omega p}$$

$$\frac{\hat{V}}{\hat{i}_c} = \left(\frac{R\sqrt{MK(M-1)}}{2M-1}\right)\frac{1}{1+s/\omega p}$$

where

$$\omega_p = \frac{2M-1}{RC(M-1)}$$

$$K = \frac{2L}{RT_s}.$$

The salient feature of the result is now apparent. Both responses show single pole response as in the duty ratio programmed converters operating in DIC mode.

An equivalent linear circuit model is developed using (6)-(8) as shown in Fig. 5 which describes completely the input and output properties of the current programmed nonlinear boost converter operating in DIC mode.

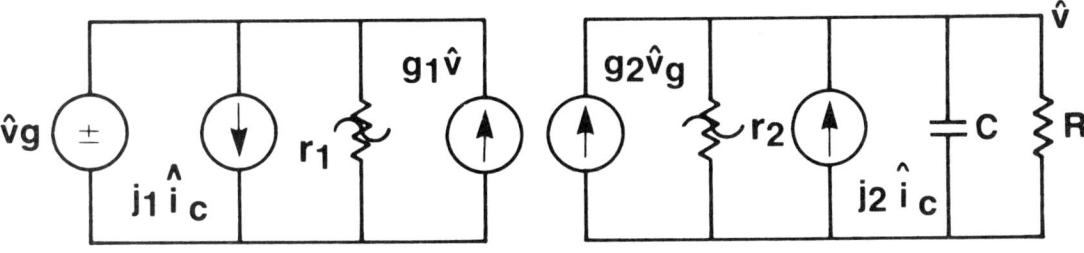

$$J_1 = M\sqrt{\frac{MK}{M-1}} \quad r_1 = \left(\frac{M-1}{M^3}\right)R \quad g_1 = \left(\frac{M}{M-1}\right)\frac{1}{R}$$

$$J_2 = \sqrt{\frac{MK}{M-1}} \quad r_2 = \left(\frac{M-1}{M}\right)R \quad g_2 = \left(\frac{M}{M-1}\right)\frac{1}{R}$$

Fig. 5. Linear equivalent circuit model for current programmed boost converter in DIC mode.

4.0 Buck and Buckboost

The same method has been followed to model the current programmed buck and buckboost converters operating in the DIC mode. The converter and its equivalent circuit are shown in Fig. 6 for the buck and in Fig. 7 for the buckboost. In both cases the switch current is the inductor current during the interval $d_1 T_s$. The equivalent circuits of Figs. 6(B) and 7(B) contain familiar current sources driving the R-C network, i.e., single pole response in both cases: the current modulation generators function of i_c, the control signal and v_g, the input modulation voltage and the filter C in parallel with load R.

The result for the two transfer functions of major interest, the line-to-output transfer function and duty ratio-to-output transfer function are as follows.

For the buck converter:

$$\frac{\hat{V}}{\hat{V}_g} = \left(\frac{M^2}{3M - 2}\right) \frac{1}{1 + s/\omega_p}$$

$$\frac{\hat{V}}{\hat{i}_c} = \left(\frac{R\sqrt{K(1 - M)}}{2 - 3M}\right) \frac{1}{1 + s/\omega_p}$$

where

$$\omega_p = \frac{2 - 3M}{RC(1 - M)}$$

$$M = \frac{V}{V_g} \quad K = \frac{2L}{RT_s}.$$

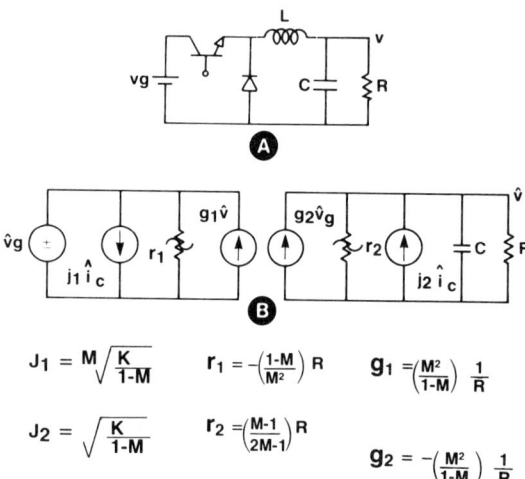

$J_1 = M\sqrt{\frac{K}{1-M}}$ $r_1 = -\left(\frac{1-M}{M^2}\right)R$ $g_1 = \left(\frac{M^2}{1-M}\right)\frac{1}{R}$

$J_2 = \sqrt{\frac{K}{1-M}}$ $r_2 = \left(\frac{M-1}{2M-1}\right)R$ $g_2 = -\left(\frac{M^2}{1-M}\right)\frac{1}{R}$

Fig. 6. (A) Buck converter. (B) Its small signal low frequency linear equivalent circuit model.

Note that as M is increased beyond two-thirds, the pole moves to the right half-plane indicating that

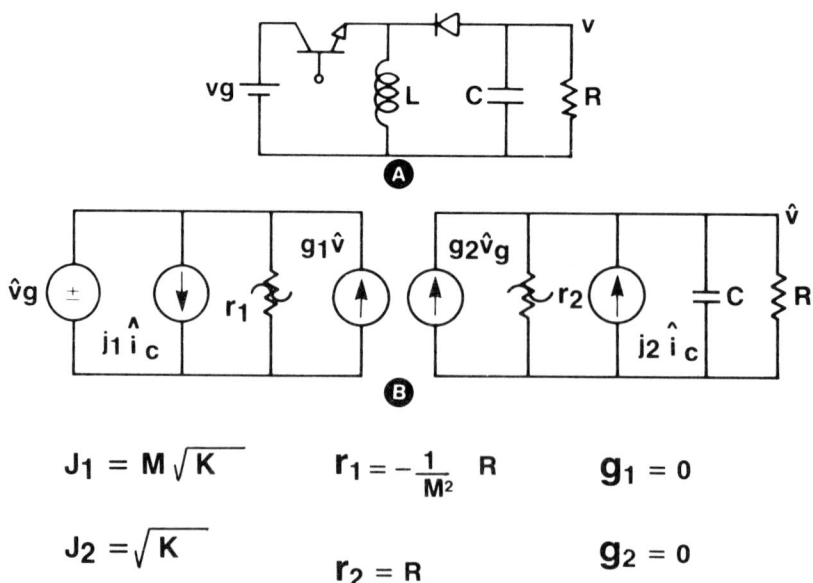

$J_1 = M\sqrt{K}$ $r_1 = -\frac{1}{M^2}R$ $g_1 = 0$

$J_2 = \sqrt{K}$ $r_2 = R$ $g_2 = 0$

Fig. 7. (A) Buckboost converter. (B) Its small signal low frequency linear equivalent circuit model.

the buck converter (open loop) becomes unstable. The physical reasoning for this unstability to occur and a remedy to avoid the same has been presented in [4].

For the buckboost converter:

$$\frac{\hat{V}}{\hat{i}_c} = \left(\frac{R\sqrt{K}}{2}\right)\frac{1}{1+s/\omega p}$$

where

$$\omega p = \frac{2}{RC}$$

$$K = \frac{2L}{RT_s}$$

It is interesting to note that $\hat{v}$ does not depend upon $\hat{v}g$ in the case of current programmed buckboost converter operating in DIC mode.

Thus, in all three of the converters, the control-to-output transfer function is a single pole response as in the duty ratio programmed converters operating in the DIC mode. As mentioned in the introduction, the reduction of the order of the system greatly simplifies the design of a regulator loop.

The results of the modeling and analysis presented above using the current injected equivalent circuit approach, i.e., 1) the transfer functions for buck, boost, buckboost are the same as those obtained by using the electronic equivalent circuit state space average approach [4] and the current injected control type approach [6], and 2) the equivalent circuit models are the same as those obtained using the electronic equivalent circuit state space average approach [4]. However, the CIECA approach presented here is more clear (compared to [5]) and is not cumbersome (compared to [7]). Also, the CIECA approach produces the linear equivalent circuit diagrams for nonlinear converters compared to current injected control theory approach (which could not produce) the equivalent circuit model. Thus, the CIECA has the merits of electronic equivalent circuit state space average approach and current injected control theory approach and devoid of the demerits.

Conclusion

The current injected equivalent circuit approach to modeling switching converters has been extended to current programmed converters operating in the DIC mode. To demonstrate the approach, the modeling is carried out for buck, boost, and buckboost converters. The analysis has revealed an instability in the buck converter (open loop) when its output-to-input voltage ratio is equal to or greater than two-thirds.

These analyses have been presented in this paper. In all the three converters, the salient feature of the model for the current programmed converters in DIC mode is that it predicts basically a one-pole response for the control-to-output transfer function. This is to be expected because of either one of these two reasons: 1) the inductor current has definite initial and final value of zero, thereby losing its state itself, or 2) in this current programmed converter the inductor receives current input rather than voltage input as in the case of duty ratio programmed converters.

Thus the modeling developed for converters in fixed frequency current programmed DIC mode and presented in this paper permits us to design regulators containing these converter power stages to achieve required performance.

References

1. P.R.K. Chetty, "Current injected equivalent circuit approach (CIECA) to modeling of switching dc-dc converters in continuous inductor conduction mode," Sundstrand Advanced Technology Corp., Rockford, IL, Internal Rep.
2. —, "Current injected equivalent circuit approach (CIECA) to modeling of switching dc-dc converters in discontinuous inductor conduction mode." Sundstrand Advanced Technology Corp., Rockford, IL, Internal Rep.
3. —, "Current injected equivalent circuit approach to modeling and analysis of current programmed switching dc-dc converters (continuous inductor conduction mode)," Sundstrand Advanced Technology Corp., Rockford, IL, Internal Rep.

4. P.R.K. Chetty and R.D. Middlebrook, "Modeling and analysis of switching dc-dc converters in current programmed discontinuous conduction mode," California Institute of Technology, Pasadena, CA, Internal Rep.
5. S. Cuk and R.D. Middlebrook, "A general unified approach to modeling switching dc-to-dc converters in discontinuous conduction mode, "*IEEE Power Specialists Conference,* 1977 Record, pp. 90-111.
6. A. Capel, M. Clique and A.J. Fossard, "Current control modulators: General theory and specific designs," *IEEE Power Electronics Specialists Conference,* 1980 Record.
7. A. Brown, "State Space Analysis of Pulsewidth Modulated Switching Converters," Power Electronics Group, California Institute of Technology, Tech, Note T-58, Feb. 6, 1979.

MODELING AND ANALYSIS OF CUK CONVERTER USING CURRENT INJECTED EQUIVALENT CIRCUIT APPROACH

The current-injected equivalent-circuit approach has been developed for modeling and analysis of switching dc-dc converters and is very versatile. This approach can also be applied for modeling and analysis of complex converters or cascaded converters. To demonstrate the ability of the current injected equivalent-circuit approach, the modeling and analysis of a Cuk converter is carried out. A small signal equivalent-circuit model is obtained which represents both input and output properties of the nonlinear converter. The results are presented in the form of linear equivalent-circuit models, as well as transfer functions.

1.0 Introduction

In the last ten years, modeling of switching dc-dc converters has received considerable attention because of the high performance requirements of power processing systems. The effort has resulted in the characterization of transfer as well as input and output properties of basically nonlinear switching dc-dc converters in the frequency domain. Among various approaches attempted to attain this goal, the current-injected equivalent-circuit approach is very versatile. This approach, having exhibited its merits [1-4], is now applied to model complex converters. The Cuk converter is modeled as an example.

Thus, this paper is concerned with the modeling and analysis of a duty ratio-programmed Cuk converter operating in fixed-frequency continuous-inductor conduction mode. The main goal of this modeling is to obtain the small signal equivalent-circuit models which represent both input and output properties. These can then be embedded in the model of a complete regulator system so that the overall dynamic properties and stability can be analyzed and designed.

The current-injected equivalent-circuit approach followed in this analysis and modeling is briefly reviewed in Section 2. Section 3 contains the detailed development of modeling for the Cuk converter. The results are compared with those obtained using the state space average electronic equivalent-circuit approach [5]. The final section presents the conclusions on the results of the work carried out.

2.0 Review

The current-injected equivalent-circuit approach to modeling switching converters in the duty ratio-programmed mode has been developed [1, 2]. Since this approach [1] is used in this paper for the modeling and analysis of converters operated in the fixed-frequency mode, a brief review of this approach is presented here.

Of course, the same approach can be also used in modeling the converters operated in the variable frequency mode.

The following conventions and notations are followed in the modeling and analysis:

$d1Ts$ interval during which the transistor is turned on and the diode is off;

$d2Ts$ interval during which the transistor is turned

© 1983 IEEE. Reprinted with permission from *IEEE Transactions on Industrial Electronics,* Vol.IE-30, No.1, pp. 56-59, Feb 1983.

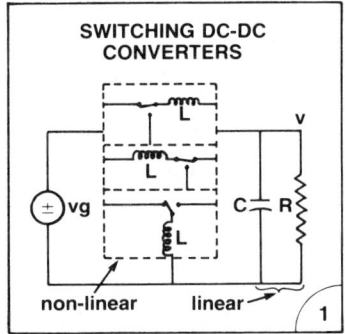
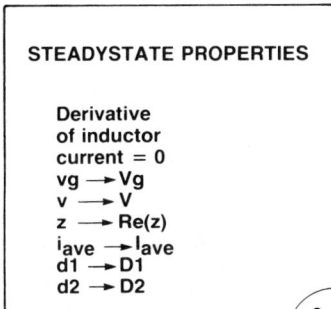
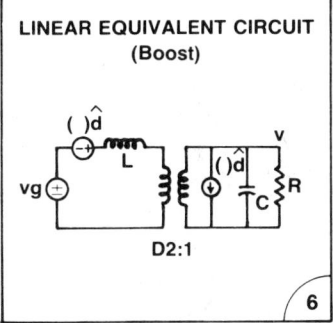

Fig. 1. Flowchart of current injected equivalent circuit approach to modeling switching dc-to-dc converters in the CIC mode.

off and the diode is on; and
$d1 Ts + d2 Ts = Ts$ and $Ts = 1/fs$
  = Switching period.

The capitilized quantities are used for steady-state values and the quantities with hats for small perturbations.

The current-injected equivalent-circuit approach to modeling converters operating in the continuous-inductor conduction mode is outlined in the flowchart of Fig. 1, which is very general, and applicable to various power stages. The first step in this process is to identify the nonlinear and linear parts of the converter circuit and linearize only the nonlinear part of the converter as the remainder of the converter is inherently linear (Box 1). The nonlinear part of the converter determines the average current injected into the linear part. Now (Box 2) a set of relationships are written referring to the converter diagram and current and voltage waveforms shown in Fig. 2, under the assumption that the corner frequency of $L\&C$ (filter components) is much smaller than the switching fre-

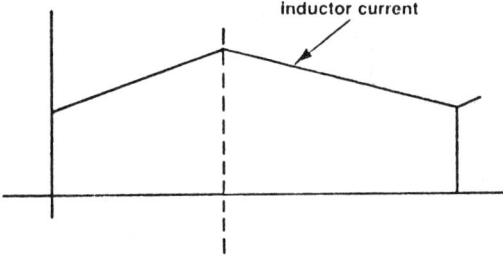
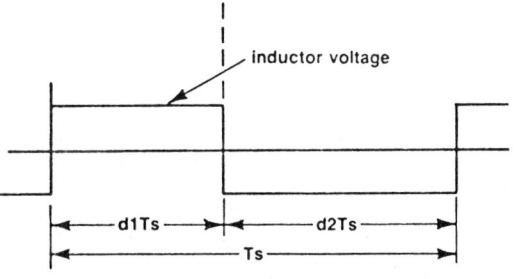

Fig. 2. Typical inductor current and voltage waveforms in buck converter.

quency (*fs*). This is true in all practical converters to achieve smaller output voltage ripple. Thus, the discontinuous nonlinear current is approximated as linear continuous current (Box 2). Now, the steady-state solution for the switching circuit is found by setting derivatives to zero (Box 3). Since the converter equations in Box 2 are linear (at a particular operating point), superposition holds and the equations can be perturbed (Box 4) by the introduction of a small ac variation over the steady-state operating point. As we know, the independent driving inputs are *vg* and *d*. The perturbation of these two inputs causes the perturbation in *i* and *v*. Now, making the small signal approximation, namely, the small ac, variations from the steady-state operating point are negligible compared to the steady-state operating point values, i.e., $\hat{v}/V$, $\hat{v}g/Vg$, $\hat{d}1/D1$, $\hat{d}2/D2$, $\hat{i}/I$ (each ≪ 1. Using the above approximations, second-order terms (product of two time-dependent quantities $\hat{d}1$ or $\hat{d}2$ and one of $\hat{V}$, $\hat{V}g$, $\hat{i}$) are neglected to obtain, once again, a linear set of equations. Now, only the ac part is retained which describes the small-signal low-frequency behavior of the converter. Using this set of equations, the input to output and control to output transfer functions (Box 5) are written. Using the same set of equations an equivalent circuit (Box 6) is drawn which represents the input and output small-signal low-frequency properties of the nonlinear converter.

3.0 Modeling of Cuk Converter

The current-injected equivalent approach to modeling converters reviewed in the previous section is applied to the Cuk converter. With the assumption of ideal switches (no parasitics or storage time modulation effects have been considered), the Cuk converter diagram is shown in Fig. 3. This converter is divided into two parts as shown in Fig. 3, for easy analysis. The first part is up to *xx* from the source *vg*, which sees an effective load of *Re* whose value is derived as the modeling and analysis progresses. A voltage of *v*1 is developed across *C*1. The second part of the converter is from *yy* to the output of the circuit.

A. First Part of the Converter. The nonlinear portion of the circuit injects a current pulse into the linear part of the circuit, i.e., *C*1 and *C*1 in turn supplies to *Re*. Inductor (*L*1) current and voltage waveforms are shown in Fig. 4. The shaded portion shows the amount of charge injected into the output linear circuit and the interval during which the current injected is *d*2*Ts*. The average inductor current injected (i_{ave1}) into the output circuit during a switching period is given by

$$i_{ave1} = d2 \cdot i1 \quad (1)$$

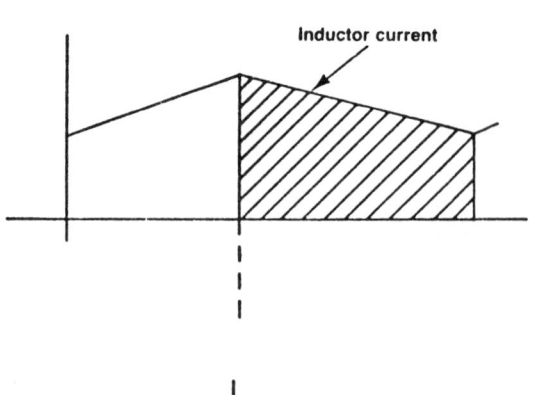

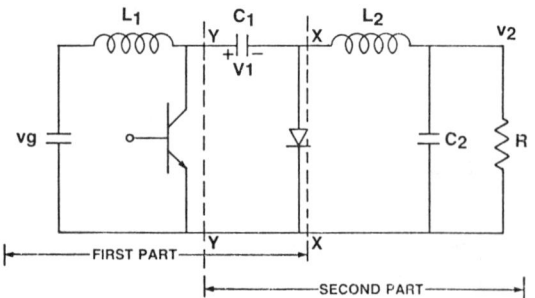

Fig. 3. Cuk converter with all parasitics and storage time effects neglected.

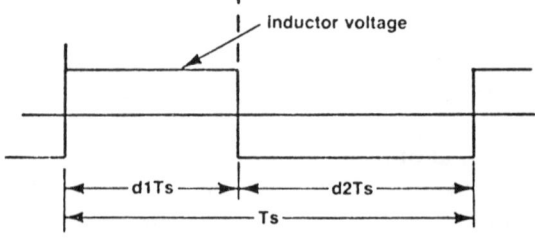

Fig. 4. Inductor (L1) current and voltage waveforms.

where $i1$ is the average inductor ($L1$) current.

The derivative of the inductor ($L1$) current is given by

$$\frac{L1 \cdot di1}{dt} = Vg - d2 \cdot V1. \quad (2)$$

The average current as expressed by (1) is injected into $C1$, which in turn is supplied to an effective load of Re. Thus, the voltage $v1$ is given by

$$v1 = i_{ave1} \cdot \left[\frac{Re}{S \cdot R3 \cdot C1 + 1} \right] \quad (3)$$

where $Re/\{1 + (S \cdot Re \cdot C1)\}$ is the impedance of the output network.

Assuming that the second part of the converter is ideal which transforms power from its input to output with an efficiency of 100 percent and noticing that the second part of the converter is, in fact, a buck converter, the value of the load seen by the first part of the converter is given by

$$Re = \frac{R}{(d1)^2} \quad (4)$$

The steady-state conditions can now be found by using (1)-(4) and setting frequency terms to zero and all other quantities to their steady-state values. Therefore, the above equations reduce to

$$\frac{V1}{Vg} = \frac{1}{D2}$$

$$I1 = \frac{V1}{Re \cdot - D2}$$

$$Re = R/D1^2. \quad (5)$$

Equations (1)-(4) are perturbed around the steady-state operating point and second-order nonlinear terms (product of two time-dependent ac variation quantities) are neglected to obtain the linear small-signal model. After taking the Laplace transform and noting that $\hat{d}2 = -\hat{d}1$

$$S \cdot L1 \cdot \hat{i}1 = \hat{V}g - D2 \cdot \hat{V}1 + V1 \cdot \hat{d}1 \quad (6)$$

$$\hat{i}_{ave1} = D2 \cdot \hat{i}1 - I1 \cdot \hat{d}1 \quad (7)$$

$$\hat{V}1 = \left[\frac{R}{D1^2 + S \cdot R \cdot C1} \right] \hat{i}_{ave1}$$

$$+ I1 \cdot D2 \left[\frac{-2R}{D1(D1^2 + SRC1)} \right] \hat{d}1. \quad (8)$$

Equations (6)-(8) can then be simplified as shown by

$$\hat{V}1 \left[1 + \frac{SL1}{R} \left(\frac{D1}{D2} \right)^2 + \frac{S^2 L1 C1}{(D2)^2} \right] = \frac{\hat{V}g}{D2} + \frac{V1}{D}$$

$$\cdot \left[1 - \frac{SL1}{R} \left(\frac{D1}{D2} \right)^2 \left(\frac{1+D2}{D1} \right) \right] \hat{d}1. \quad (9)$$

B. Second Part of the Converter. The converter circuit from yy to the output is considered here. One can see that this is obviously a buck converter with $v1$ being the input voltage. Inductor ($L2$) current and voltage waveforms for the second part of the converter are shown in Fig. 5. The shaded portion shows the amount of charge injected into the output linear circuit and the interval during which the current injected is T_s. The average

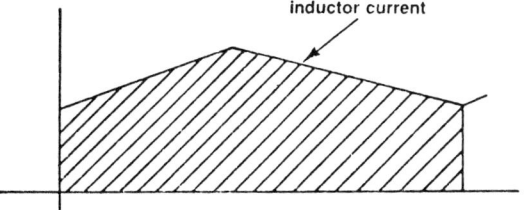

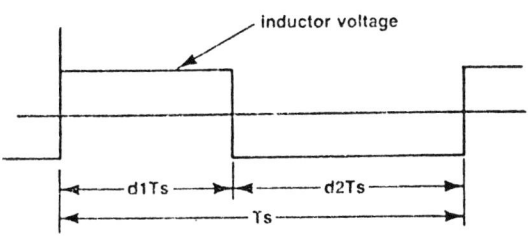

Fig. 5. Inductor ($L2$) current and voltage waveforms.

inductor (L2) current injected (i_{ave2}) into the output circuit during a switching period is given by

$$i_{ave2} = i2 \quad (10)$$

where $i2$ is the average inductor current.

The derivative of the inductor (L2) current is given by

$$L2 \cdot \frac{di2}{dt} = V1 \cdot d1 - V2 \quad (11)$$

where $V2$ is the output voltage.

The voltage $v2$ is given by

$$v2 = i_{ave2}\left[\frac{R}{1 + SRC2}\right]. \quad (12)$$

The steady-state conditions can now be found by using (10)-(12) and setting frequency terms to zero and all other quantities to their steady-state values. Therefore, the above equations reduce to

$$\frac{V2}{V1} = D1$$
$$I2 = \frac{V2}{R}. \quad (13)$$

Equations (10)-(12) are perturbed around the steady-state operating point, and second-order nonlinear terms are neglected once again to obtain the linear small-signal model. After taking the Laplace transform.

$$\hat{i}_{ave2} = \hat{i2} \quad (14)$$

$$S \cdot L2 \cdot \hat{i2} = D1 \cdot \hat{V1} + V1 \cdot \hat{d1} - \hat{V2} \quad (15)$$

$$\hat{V2} = \hat{i}_{ave2}\left[\frac{R}{1 + S \cdot R \cdot C2}\right] \quad (16)$$

(14)-(16) are simplified as shown by

$$\left[1 + \frac{S \cdot L2}{R} + S^2 L2 C2\right]\hat{V2} = D1 \cdot \hat{V1} + V1 \cdot \hat{d1}. \quad (17)$$

Now, (9) and (17) are combined to get the overall input to output and control to output transfer functions which are given by

$$\frac{\hat{V2}}{\hat{Vg}} = \left(\frac{D1}{D2}\right)$$
$$\frac{1}{\left[+\frac{S}{W1Q1}+\frac{S^2}{W2^2}\right]\left[1+\frac{S}{W2Q2}+\frac{S^2}{W2^2}\right]}$$

$$\frac{\hat{V2}}{\hat{d}} = \left(\frac{V2}{D1 \cdot D2}\right)$$

$$\frac{\left[1 - \frac{S}{QZWz} + \frac{S^2}{Wz^2}\right]}{\left[1 + \frac{S}{W1Q1} + \frac{S^2}{W1^2}\right]\left[1 + \frac{S}{W2Q2} + \frac{S^2}{W2^2}\right]}$$

(18)

where

$$W1^2 = \frac{D2^2}{L1 \cdot C1}$$

$$W2^2 = \frac{1}{L2 \cdot C2}$$

$$W_z^2 = \frac{D2}{L1 \cdot C1}$$

$$\frac{1}{W1Q1} = \frac{1}{WzQz} = \frac{L1}{R}\left(\frac{D1}{D2}\right)^2$$

$$\frac{1}{W2Q2} = \frac{L2}{R}.$$

An equivalent linear circuit model is developed using (6)-(8), (14)-(16) as shown in Fig. 6, which

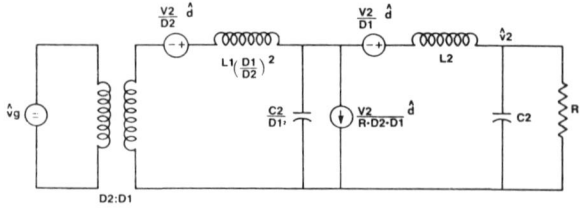

Fig. 6. Linear equivalent circuit model for Cuk converter in CIC mode.

describes completely the input and output properties of the duty ratio-programmed nonlinear Cuk converter operating in the continuous inductor conduction mode.

4.0 Conclusions

The current-injected equivalent-circuit approach has been applied to the modeling of Cuk converter to demonstrate its ability to easily model even complex converters or cascaded converters. One can see that the modeling and analysis is very simple. The results of the modeling are the same as those obtained using electronic equivalent-circuit state space average approach. The two approaches are essentially the same and it is primarily the averaging that is done differently.

Thus, the modeling developed for a duty ratio programmed Cuk converter operating in the fixed-frequency continuous-inductor conduction mode and presented in this paper permits us to design regulators containing Cuk converter power stages to achieve the desired performance.

References

1. P.R.K. Chetty, "Current injected equivalent circuit approach (CIECA) to modeling of switching dc-dc converters in continuous inductor conduction mode," *IEEE Trans. Aerosp. Electron. Syst.,* vol. AES-17, no. 6, Nov. 1981.
2. P.R.K. Chetty, "Current injected equivalent circuit approach (CIECA) to modeling of switching dc-dc converters in discontinuous inductor conduction mode," *IEEE Trans. Ind. Electron.,* vol IE-29, no. 3, Aug. 1982.
3. P.R.K. Chetty, "Current injected equivalent circuit approach to the modeling of current programmed switching dc-dc converters," Internal Rep., accepted for publication in *IEEE Trans. Aerospace and Electronic Systems.*
4. P.R.K. Chetty, "Current injected equivalent circuit approach to the modeling and analysis of current programmed switching dc-dc converters (Discontinuous inductor conduction mode)," *IEEE Trans. Ind. Applicat.,* vol. IA-18, no. 3, May/June 1982.
5. Slobodan Cuk and R.D. Middlebrook, "A new optimum topology switching dc-to-dc converter," in *Rec. 1977 IEEE Power Electronics Specialists Conf.,* pp. 160-179.

Chapter 3

Design and Measurements

Modeling and Design of Switching Regulators 48

Closed Loops--on Track for Testing Switchers 60

Measurement of Magnitude and Phase of Switching Regulator Transfer Functions and Loop Gain 69

MODELING AND DESIGN OF SWITCHING REGULATORS

Various building blocks of a switching regulator are described in detail and mathematical models are developed for all building blocks in terms of transfer functions, which enable one to design a switching regulator for stability, desirable bandwidth, line rejection, and transient response. A step-by-step procedure to design compensation is illustrated using two examples. Various networks for compensation and their transfer functions are presented which the author hopes will be very handy to use and will become the reference source.

1.0 Introduction

In the last ten years, switching regulators have received considerable attention because of the high performance requirements of power processing systems. Switching mode regulators have almost replaced the conventional dissipative series regulators because of their inherent superior characteristics, i.e., high efficiency, small size and weight, low volume, low weight, and equal reliability. These regulators are very useful in dc-to-dc, dc-to-ac, and ac-to-ac conversions and to buck or boost the voltage levels with isolation. Switching mode regulators have to be well understood before one can aim for optimum performance. Modeling and analysis of switching mode regulators is very important, but the design of switching regulators for stability, desirable bandwidth, better transient response, and better line rejection has not had much coverage in the literature. Hence, various building blocks of switching regulators are described, mathematical models are developed for each building block, and finally a complete model for a switching regulator is obtained. A step-by-step procedure for designing the compensation is described and two examples are given as an illustration. The compensation is realized using appropriate networks. Various networks for compensation are also presented along with their transfer functions.

Thus the main purpose of this paper is the modeling and design of switching regulators. Section 2 describes switching mode regulators. Section 3 presents the modeling of the various building blocks. Section 4 gives stability criteria, a step-by-step procedure for designing the compensation, two illustrative examples, and different networks for compensation and their transfer functions.

2.0 Switching Regulators

Switching regulators operate on the principle of storing energy in an inductor during one portion of the cycle and then transferring the stored inductive energy to a capacitor in another portion of the cycle. This is in contrast to the series dissipative regulators where, to keep the output constant, the difference voltage between input and output is dropped across a variable resistor (a transistor in linear or conduction mode of operation). As the transistor in a switching regulator is operated either in saturation or cutoff (and ideally inductor and capacitor are lossless), the switching regulators possess high efficiency.

A block schematic of a typical switching regulator is shown in Fig. 1. This consists of switching dc-to-dc converter which is the power stage, a voltage divider network, a stable voltage reference, an error amplifier, a compensation network, a pulsewidth modulator, and a driver stage. The reduced output voltage is compared with the reference; an error signal is amplified and fed into the pulsewidth modulator which drives the power stage to determine the required output quantities. Having already established the small signal low frequency model for the power stage [1], other building blocks now remain for modeling.

A. Power Stage or Switching Dc-Dc Converter. Switching dc-dc converters are the power stages for switching regulators. Buck, boost, and buck-boost converters are the basic fundamental converters (Fig. 2). All other converters (Fig. 3) are derived from these basic converters. For example, the forward converter [Fig. 3(A)] is the buck converter with input-output isolation and the quasi-square wave push-pull converter [Fig. 3(B)] is the

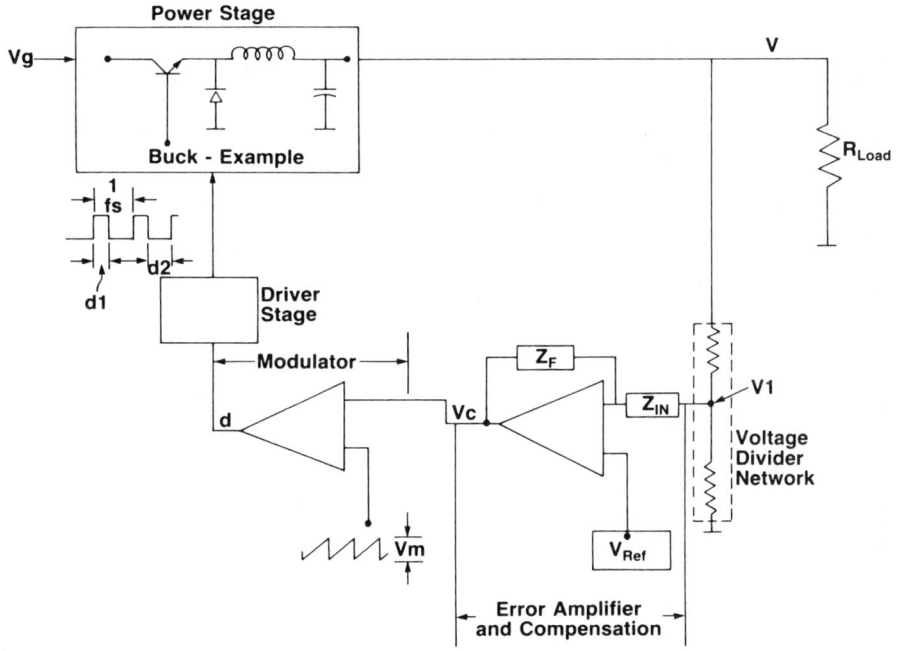

Fig. 1. Typical switching regulator.

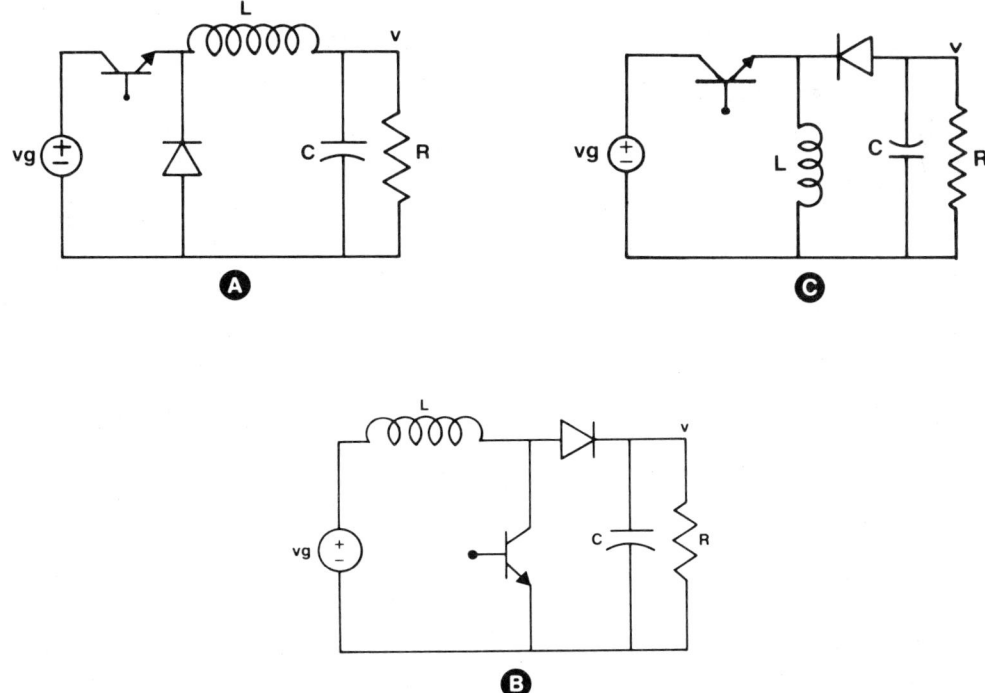

Fig. 2. Buck, boost, buckboost—basic converters.

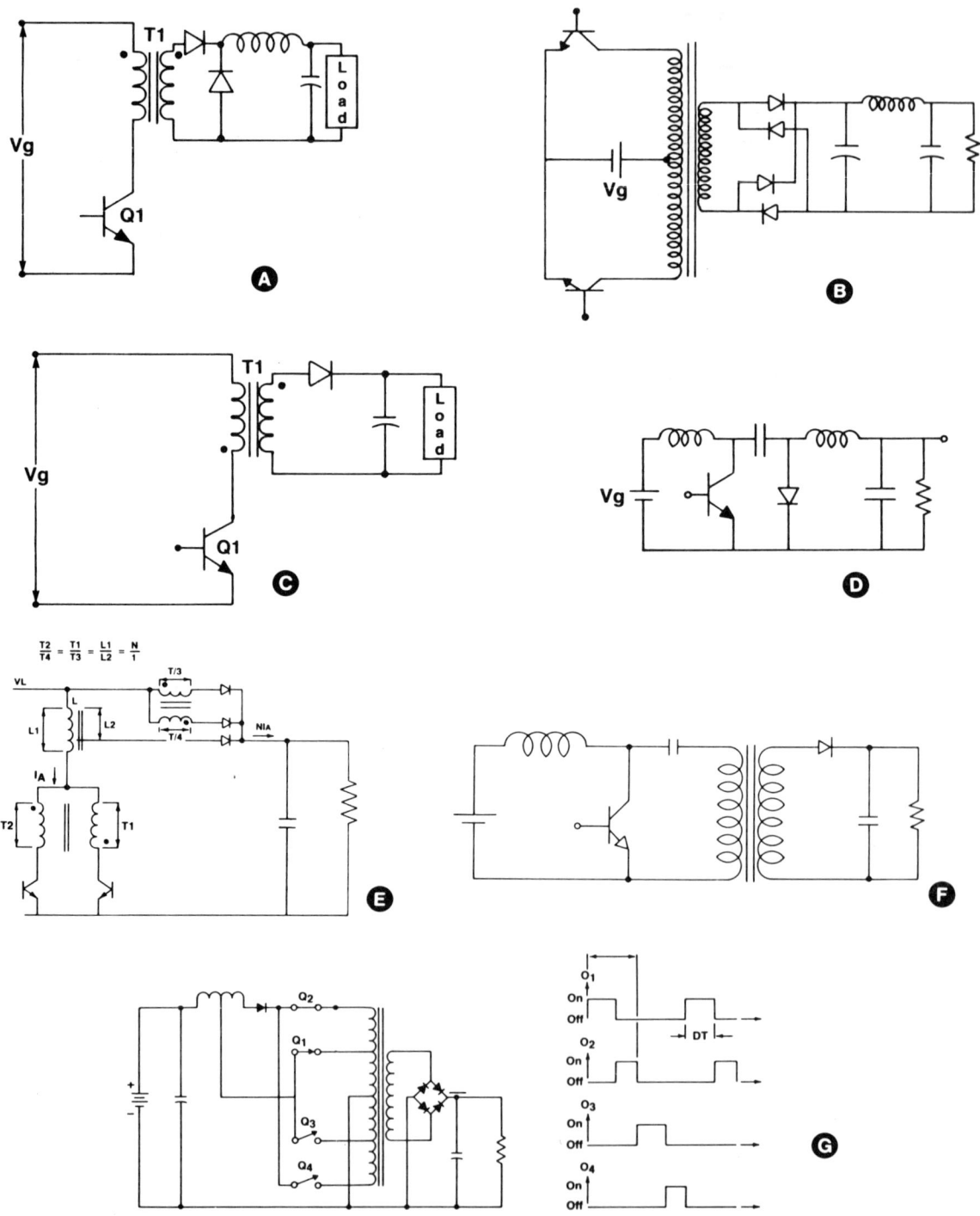

Fig. 3. Converter extensions (A) Forward converter (B) Push-pull quasi-square wave converter (C) Flyback converter (D) Cuk converter (E) Weinberg converter (F) Bell Lab converter (G) Venable converter.

two-phase forward converter derived to share the power handling capacity by two power transistor switches instead of one. The flyback converter [Fig. 3(C)] is the buck-boost converter with input-output isolation. The Cuk converter [Fig. 3(D)] is a simplified cascaded boost-buck converter utilizing a minimum number of switches. Basically these converters are nonlinear.

3.0 Modeling

For several years, modeling and analysis of such nonlinear switching dc-dc converters has been carried out. This means that a partial solution to the problem is already available. Among the different approaches to modeling switching dc-dc converters is the current injected equivalent circuit approach (CIECA)[1] which is briefly reviewed here because of its merits over the other approaches.

A. Power Stages or Switching Dc-Dc Converters.

The following conventions and notations are followed in the modeling and analysis:

$d_1 T_s$ interval during which the transistor is turned on and the diode is off;

$d_2 T_s$ interval during which the transistor is turned off and the diode is on;

$d_1 T_s + d_2 T_s = T_s$ and, $T_s = 1/f_s$ is the switching period.

In the discussion, the capitalized quantities are used for steady state values and the quantities with carets are used for small perturbations.

Modeling converters operating in continuous inductor conduction mode according to CIECA is outlined in the flowchart of Fig. 4, which is very general and applicable to various power stages. The first step in this process is to identify the nonlinear and linear parts of the converter circuit and to linearize only the nonlinear part of the converter

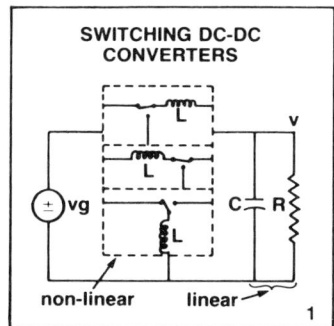
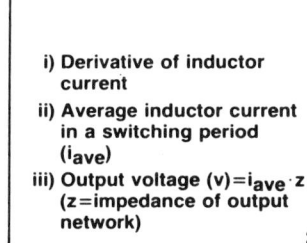
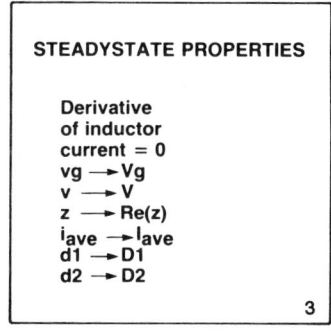
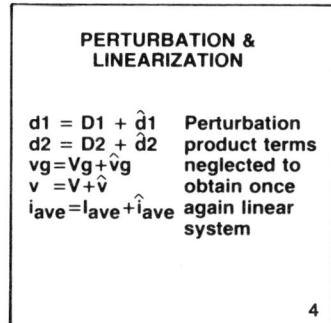
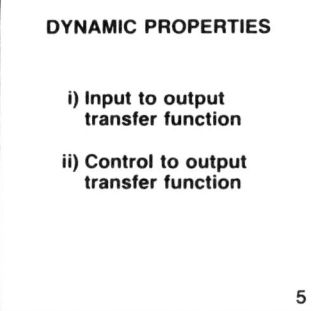
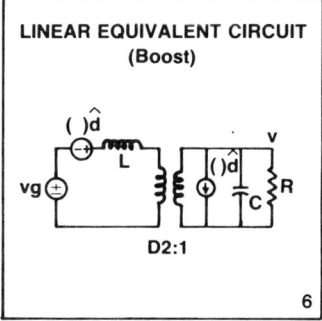

Fig. 4. Flowchart illustrating modeling of switching dc-dc converters using current injected equivalent circuit approach (CIECA).

as the remainder of the converter is inherently linear (box 1). The nonlinear part of the converter determines the average current injected into the linear part. Now (box 2) a set of relationships are written referring to the converter diagram and current and voltage waveforms shown in Fig. 5.

1) average current (i_{ave}) injected into the linear part in a switching period;
2) derivative of the inductor current function of the value of the inductor, the voltage across that in a switching period;
3) relationship between average injected current and output voltage $v = i_{ave} z$, where z is the impedance of the linear part of the converter.

Now steady state solution is achieved by setting derivatives and perturbations to zero (box 3). Since the converter equations in box 2 are linear, superposition holds and can be perturbed (box 4) by the introduction of a small ac variation over the steady state operating point. As we know, the independent driving inputs are vg and d, and the perturbation in these two inputs causes the perturbation in i and v. Now making the small signal approximation, namely, the small ac variation from the steady state operating point values, i.e., $\hat{v}/V$, $\hat{v}g/Vg$, $\hat{d}_1/D_1$, $\hat{d}_2/D_2$, $\hat{i}/I$ (each) ≤ 1. Using the above approximations, nonlinear second order terms are neglected to obtain the linear set of equa-

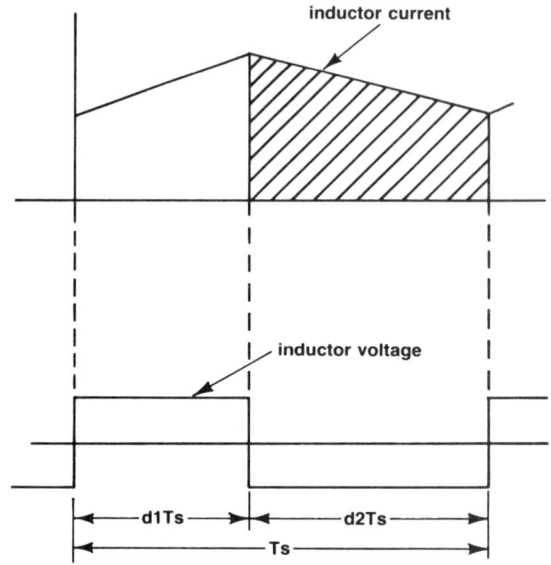

Fig. 5. Typical inductor voltage and current waveforms in switching dc-dc converter.

tions again. Only the ac part is retained, which describes the small signal, low frequency behavior of the converter. Using this set of equations, the input-to-output and control-to-output transfer functions (box 5) are written. Using the same set of equations, an equivalent circuit (box 6) is drawn which represents the input and output small signal, low frequency properties of the nonlinear converter.

B. Modeling of Buck, Boost, and Buck-Boost Converters. Modeling converters accor-

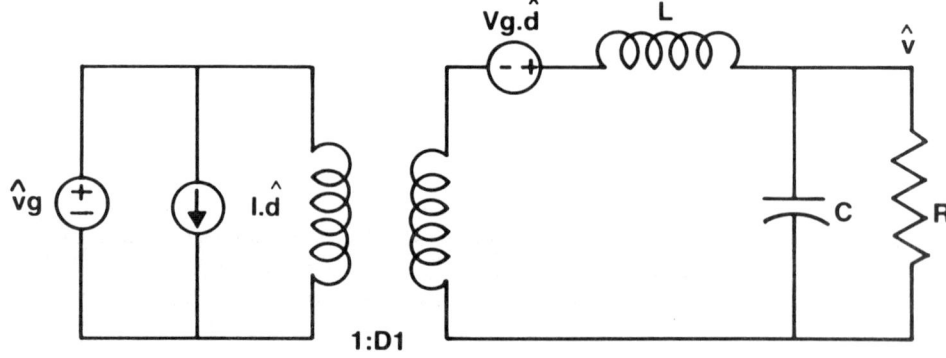

Fig. 6. Low frequency small signal equivalent circuit model for buck converter.

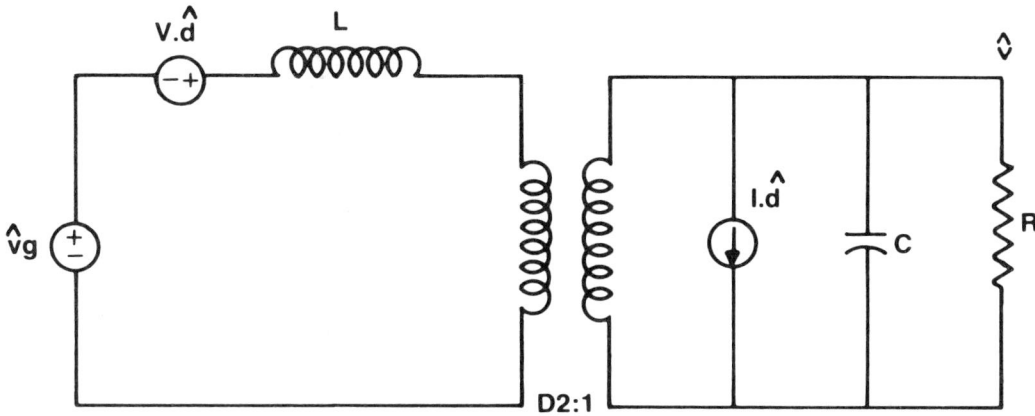

Fig. 7. Low frequency small signal equivalent circuit model for boost converter.

ding to CIECA is reviewed above and applied to buck, boost, and buck-boost converters, and the important results, i.e., input-to-output and duty ratio-to-output transfer functions, are presented here along with their small signal, low frequency equivalent circuits (Figs. 6-8).

Buck Converter:

$\hat{V}(s)/\hat{V}g(s) = D_1 (1 + SL/R + S^2LC)^{-1}$

$\hat{V}(s)/\hat{d}_1(s) = (V/D) (1 + SL/R + S^2LC)^{-1}$.

Boost Converter:

$\hat{V}(s)/\hat{V}g(s) = D_2^{-1} (1 + SL/R\, D_2^2 + S^2LC/D_2^2)^{-1}$

$\hat{V}(s)/\hat{d}_1(s) = (V/D_2) [(1 - SL/R\, D_2^2)/(1 + SL/R\, D_2^2 + S^2LC/D_2^2)]$.

Buck-boost Converter:

$\hat{V}(s)/\hat{V}g(s) = D_1/D_2 (1 + SL/R\, D_2^2 + S^2LC/D_2^2)^{-1}$

$\hat{V}(s)/\hat{d}_1(s) = (V/D_1\, D_2) (1 - SL/R\, D_2^2) / (1 + SL/R\, D_2^2 + S^2LC/D_2^2)$.

C. Error Amplifier and Compensation.

Figure 9 shows a typical error amplifier and compensation network. The reduced output voltage is compared with a stable reference and the error volt-

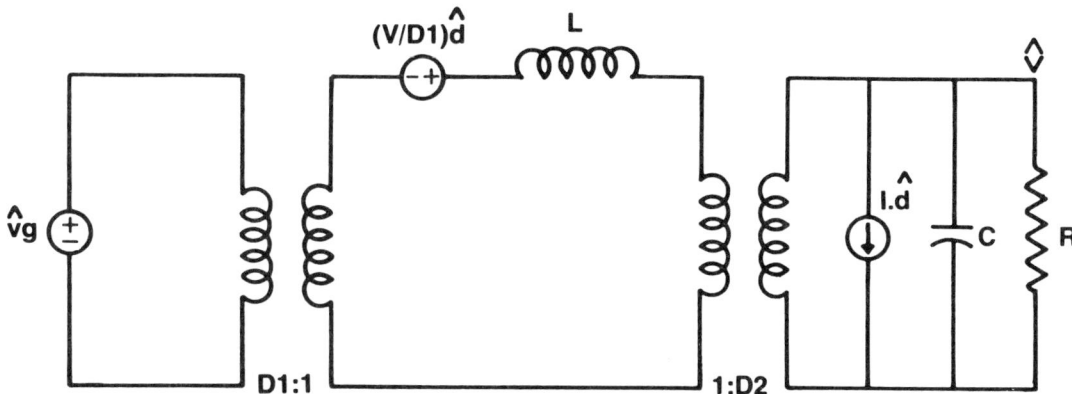

Fig. 8. Low frequency small signal equivalent circuit model for buckboost converter.

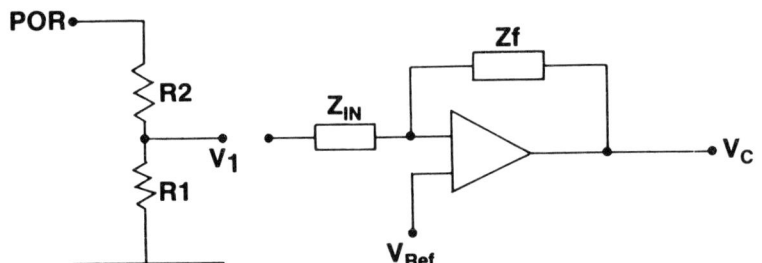

Fig. 9. Voltage divider network and error amplifier.

age is amplified with proper compensation. Since the circuit is linear, the transfer function can be written easily as given below. Depending upon the power stage and other requirements, the compensation has to be designed.

For a voltage divider network,

$$\hat{V}_1(s)/\hat{V}(s) = R_1/(R_1 + R_2) = K.$$

For an error amplifier,

$$\hat{V}_c(s)/\hat{V}_1(s) = Z_f/Z_{in} = A(s).$$

D. Pulsewidth Modulator. The pulsewidth modulator converts an analog control voltage into a duty ratio which drives the switch. Figure 10(A) shows a typical pulsewidth modulator. The amplified error voltage is compared with a sawtooth waveform and a pulsewidth modulated signal is produced. From the waveform shown in the figure, it can be seen that if the amplified error voltage is equal to the height of the sawtooth ramp, then the duty ratio is 100 percent. Thus the transfer function is given by

$$\hat{d}_1(s)/\hat{V}_c(s) = V_m^{-1} = H_m(S).$$

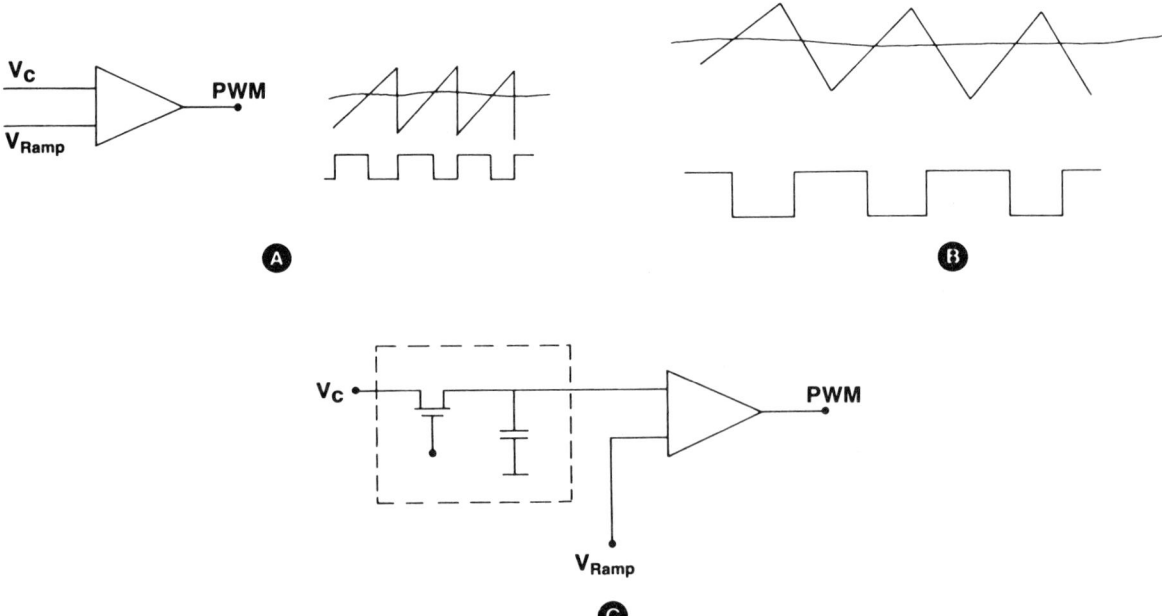

Fig. 10. Typical pulse width modulator (A) Synchronous/Single ramp (B) Triangular ramp (C) With sample and hold.

There are other pulsewidth modulators used in specific circumstances for various reasons. A pulsewidth modulator followed by a sample and hold [Fig. 10(C)] is employed if the signal is too noisy or varying fast.

Figure 1 is redrawn as shown in Fig. 11 by removing the modulator and showing its effect on the voltage and current-dependent generators directly. Thus Fig. 11 shows the general small signal ac equivalent linear circuit model for the nonlinear switching mode regulator. The principal performance specifications of a regulator are concerned with its dc regulation, output impedance, transient response, and line rejection (audio susceptibility). All these properties are closely related to the regulator loop gain. The loop gain is simply the product of the gains of all the building blocks and the phase shift is the sum of the phase shifts of all the building blocks. The loop gain and its phase dictates the stability against oscillation. Thus all these properties are determined by the various building blocks of switching regulators.

The loop gain T is given by

$$T = [\hat{V}(s)/\hat{d}_1(s)] \, K \, A(s) \, H_m(s). \quad (1)$$

For the converter shown in Fig. 1,

$$T = (V/D)(1 + SL/R + S^2LC)^{-1} [R_1/(R_1 + R_2)] \, V_m^{-1} A(s). \quad (2)$$

Input-to-output transfer function (line rejection) (open loop) for the buck converter example is given by

$$F = D_1 (1 + SL/R + S^2LC)^{-1}. \quad (3)$$

When the loop is closed around the converter,

$$F_{(closed\ loop)} = F_{(open\ loop)}/(1 + T). \quad (4)$$

4.0 Stability Criteria

For a system to be stable, the loop gain must fall below unity by the time the total phase shift has

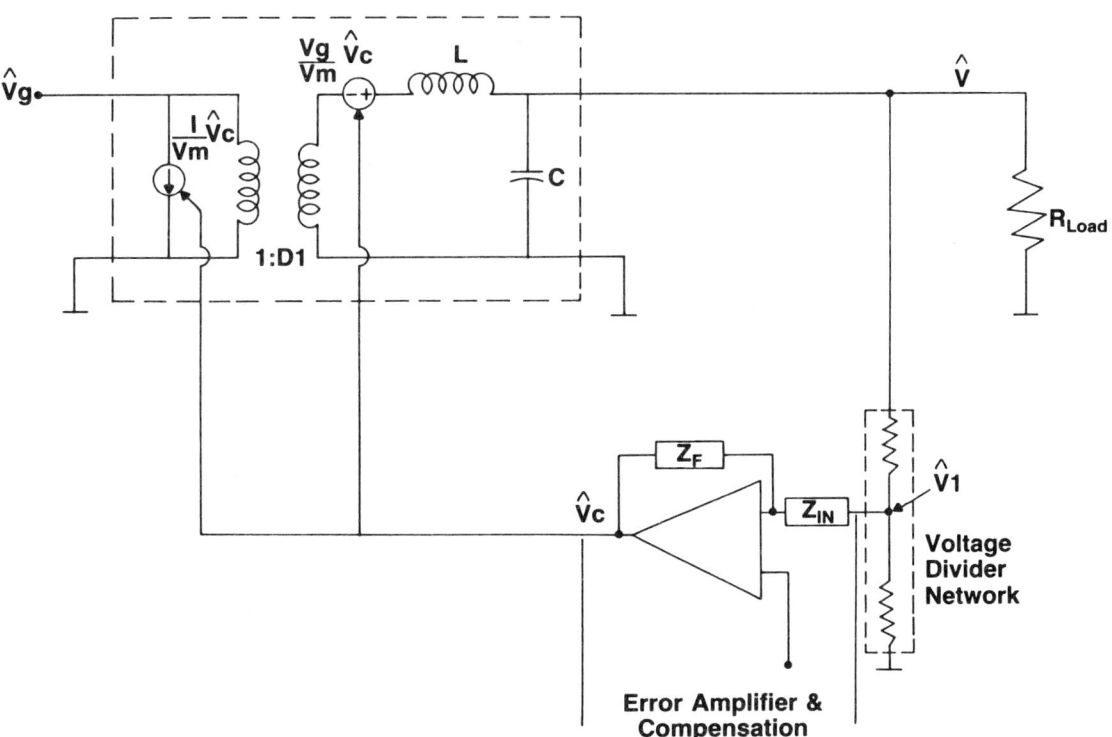

Fig. 11. Low frequency small signal equivalent circuit model for switching regulator.

reached 360 deg. The gain margin is defined as the amount of gain below unity when the total phase shift is 360 deg. The phase margin is defined as the difference between the actual phase shift when the loop gain is unity and 360 deg. (Fig. 12). Stability is sometimes described in terms of 180 deg. of phase shift. This is because even at dc, the feedback is negative, i.e., there is a phase inversion of 180 deg.

Now a step-by-step procedure for designing the compensation is presented with two examples.

A. Example 1. Equation (1) gives the loop gain of the regulator. Now let us start, assuming that the gain of the error amplifier is 1 and that it does not contain any frequency terms. The compensation is designed at the end.

The dc gain is given by

$$dc\ gain = (V/D)\ [R_1/(R_1 + R_2)]\ V_m^{-1}.$$

Because of the power stage, there is -40 dB slope starting at a frequency equal to the corner frequency of the power stage filter. This is shown in Fig. 13(A). The value of the load resistance, parasitics accompanied with L and C, and characteristic resistance of the filter determine the damping of the filter (value of the Q-factor) [Fig. 13(B)]. A 0 dB line is placed as desired by properly selecting the component values in the dc gain term. Say x dB down in the 0 dB line. A double-sided arrow mark across the 0 dB line (Fig. 13C) shows that there is flexibility to move the 0 dB line up or down, either by selecting a dc gain term or by adjusting the gain of the error amplifier to the required level. The system oscillates if the loop is closed without any compensation as the phase shift is 180 deg. at 0 dB crossover. Therefore a pole is placed as shown in Fig. 13(D). Again, the double-sided arrow mark across -20 dB slope line indicates that, depending upon the pole frequency f_p, it can be moved to left or right. If it is moved to the left, the bandwidth will be less but the system will be stable. However, since the bandwidth is small, the transient response will be poor. Hence the pole is moved towards the right side so that it cuts the 0 dB line at a frequency f_{p0} less than the corner frequency of the power stage filter f_{1c} as shown in Fig. 13(E). The system will be stable as the loop gain crosses the 0 dB line with -20 dB slope. The gain and phase margins can be determined easily. The transient response of the system will be relatively better as the bandwidth has increased. However, a further increase in bandwidth is not possible in this approach because the corner frequency of the power stage filter occurs in the neighborhood of the f_{p0} which immediately adds a 90 deg. phase shift at f_{1c}. An

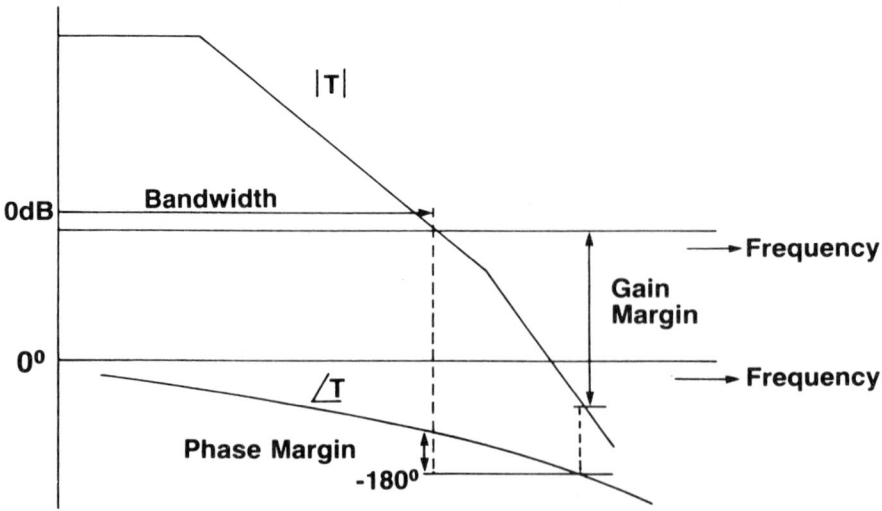

Fig. 12. Stability definitions—gain margin, phase margin, and bandwidth.

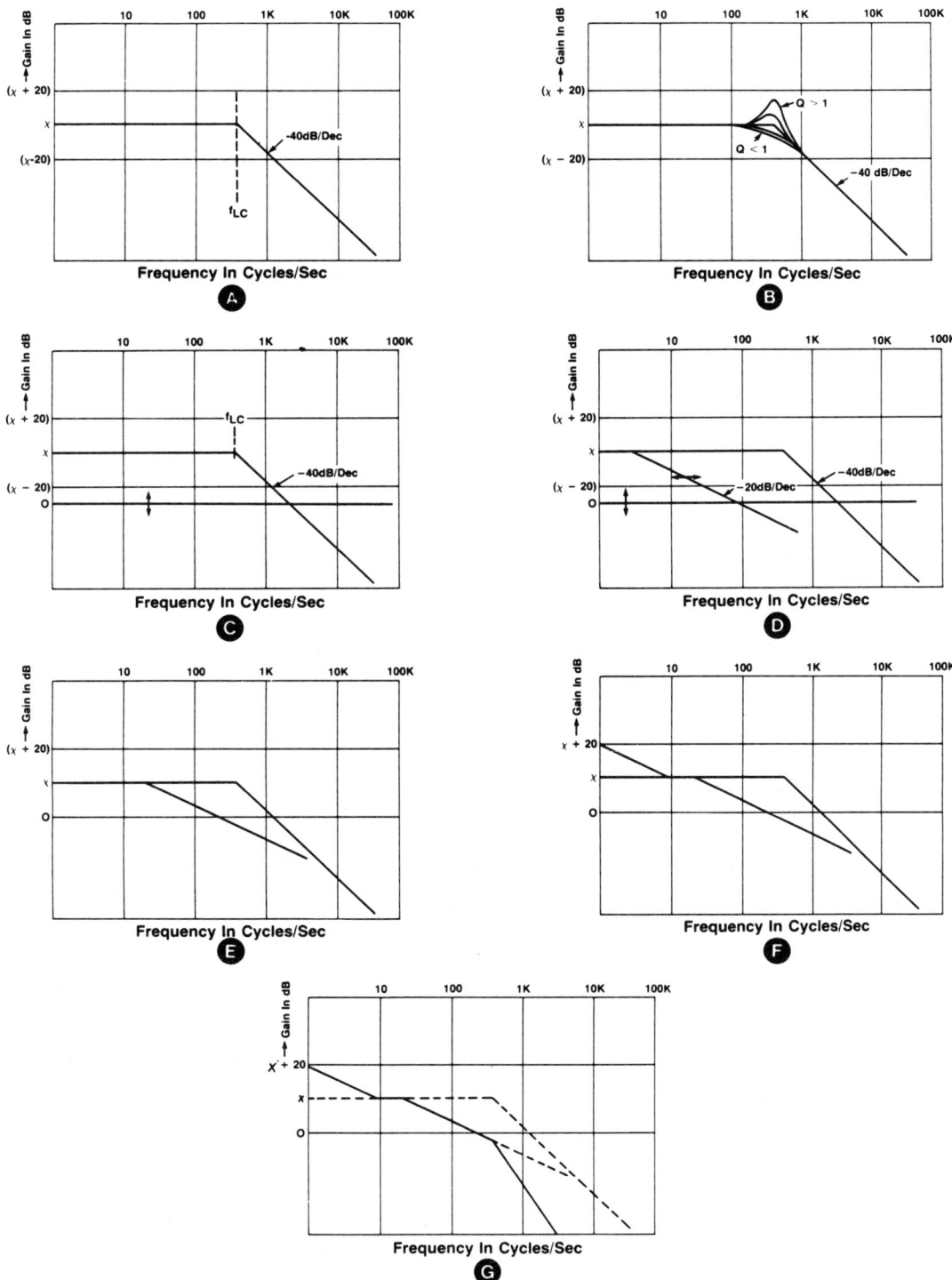

Fig. 13. Gain versus frequency—Compensation for example 1.

alternative approach is discussed in the second example. A further improvement in the loop gain is achieved by placing a pole-zero such that the zero f_z frequency is lower than the f_p [Fig. 13(F)]. This increases the low frequency gain, which results in improved line regulation and load regulation. Figure 13(G) gives the loop gain including complete compensation.

The compensation is realized using practical circuits. From experience, the dominant pole is achieved by adding a capacitor appropriately to the voltage divider network and pole-zero by modifying the error amplifier as shown in Fig. 14. This transfer function is given by

$$\hat{V}_1/\hat{V}_c = [R_A/(R_A + R_B)][1 + SC_1(R_A \| R_B)]^{-1}$$
$$\hat{V}_c/\hat{V}_1 = (1 + SR_2C_2)/SR_1C_2.$$

Such a compensation has been successfully implemented for a 2 kW quasi-square wave push-pull converter regulator.

B. Example 2. As mentioned in Example 1, the achievable bandwidth is limited by the corner frequency of the power stage filter. Further increase in bandwidth is attempted here. Again start with Fig. 13(A). Now a zero is placed as shown in

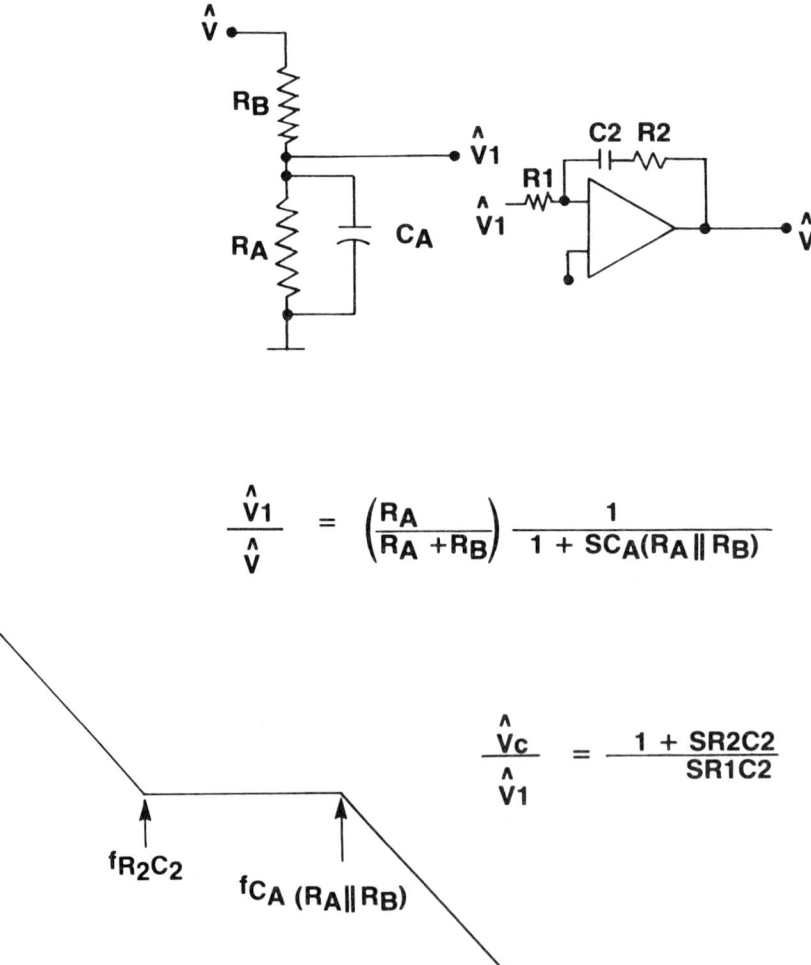

Fig. 14. Compensation network for example 1.

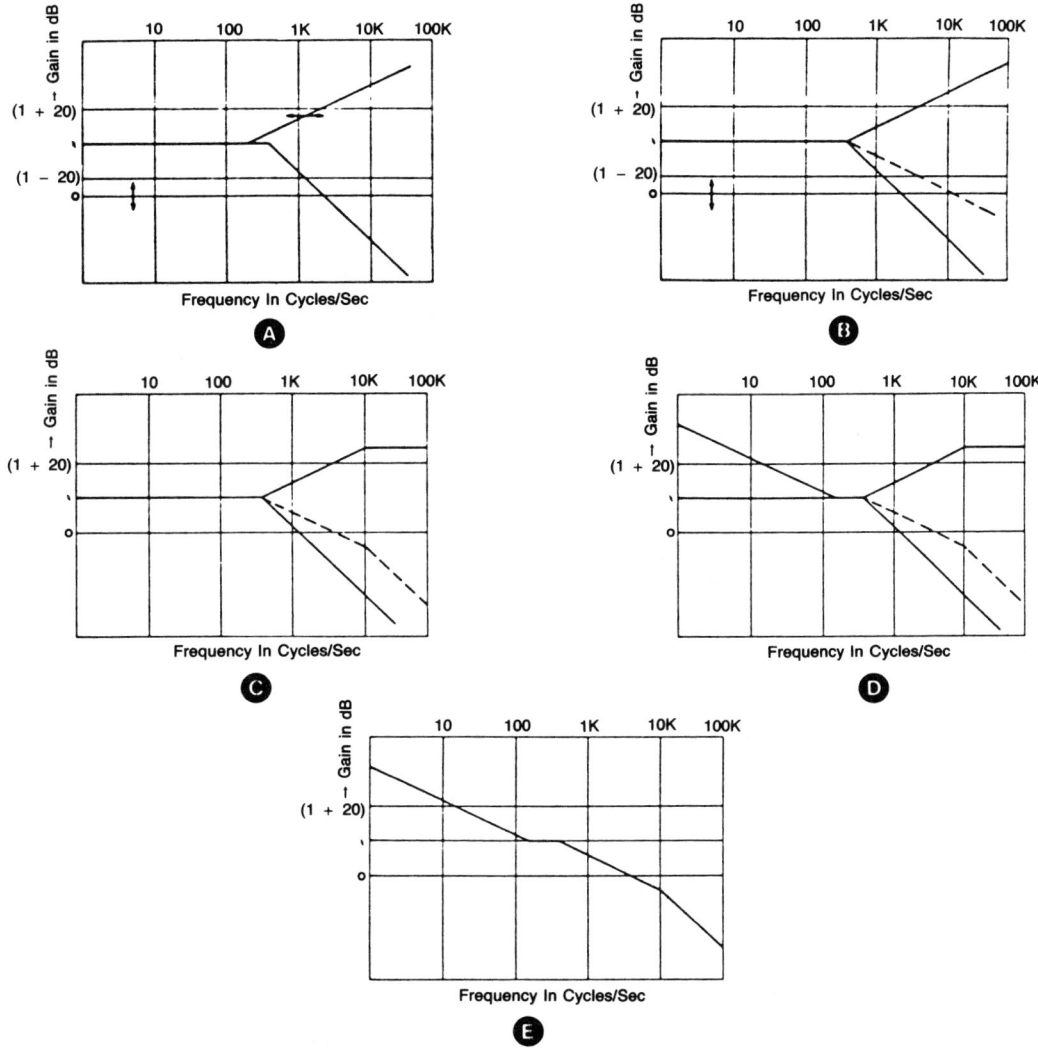

Fig. 15. Gain versus frequency—Compensation for example 2.

Fig. 15(A). Depending upon the zero frequency f_{z1}, the +20 dB slope line can be moved toward the left or the right. Selecting zero frequency f_{z1} to be equal to the corner frequency of the power stage filter f_{1c} results in a single pole (−20 dB slope) as shown in Fig. 15(B). Adjusting the gain, the 0 dB line is placed such that the required bandwidth is achieved. It is better not to allow high frequency components after zero crossover and hence a pole is added to the system at a frequency higher than the zero crossover frequency [Fig. 15(C)]. Further improvement in the loop gain is achieved as in Example 1 by placing a pole-zero such that the zero (f_{z2}) frequency is lower than f_{1c} [Fig. 15(D)]. This increases the low frequency gain which results in improved line regulation and load regulation. Figure 15(E) gives the look gain including complete compensation. The compensation is realized using the circuit shown in Fig. 16, whose transfer function is

$$\hat{V}_c/\hat{V} = [(1 + SR_1C_1)/SC_2R_1R_{11}] \{(1 + SR_2C_2)/[1 + SC_1(R_1 \| R_{11})]\}.$$

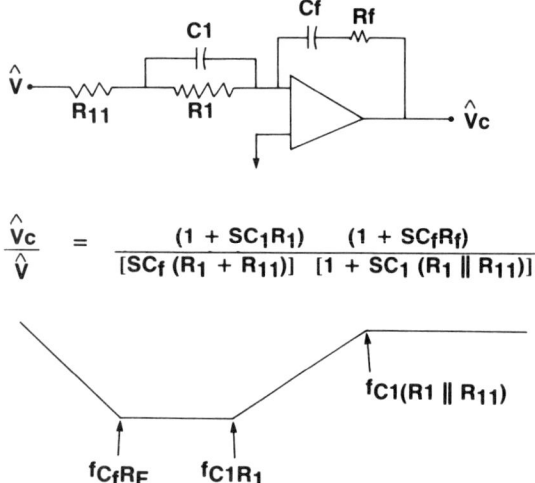

Fig. 16. Compensation network for example 2.

Such a compensation has been successfully implemented for a buck converter-regulator with a peak power capability of 10 kW.

These two examples give a good idea about designing proper compensation to achieve stable regulator operation with a large bandwidth and gain to meet transient response and line rejection requirements.

Though the circuits to realize the compensation have been rightly selected at once in the above examples, in practice it requires some experience.

To make such circuits handy, various networks for compensation and their transfer functions are given in Table 1. The author hopes that this information will be helpful as a reference to realize various compensation schemes.

Conclusions

The object of this paper has been to model the complete switching regulator and to present various networks for compensation which are handy to use.

After a brief description of switching regulators in Section 2, modeling of switching regulator building blocks was presented in Section 3. A step-by-step procedure for designing the compensation was presented with two examples in Section 4. Such designs have been successfully implemented on switching regulators at the 2 kW to 10 kW level. Various networks for compensation and their transfer functions were presented for easy reference.

References

1. Chetty, P.R.K., (1981) Current injected equivalent circuit approach (CIECA) to modeling of switching dc-dc converters in continuous inductor conduction mode. *IEEE Transactions on Aerospace Electronics and Systems,* Nov. 1981, AES-17, 802-808.

CLOSED LOOPS—ON TRACK FOR TESTING SWITCHERS

Measuring frequency response is the best way to characterize a switching-mode power supply. A closed-loop approach saves time and money by using a standard current probe.

Probably the most difficult task facing today's OEM designer of switching-mode power supplies is checking out a completed unit to verify whether it meets the design specifications. Experience shows that a supply's basic characteristics can be best determined by measuring frequency response, gain, and phase, preferably when the supply is in its normal configuration—that is, a closed loop. Closed-loop frequency measurements, however, usually require special, and costly, equipment because these systems are inherently nonlinear and produce noise in the switching process. But a new approach that employs a standard current probe simplifies the testing, as well as reducing both time and cost. With this approach, characteristics such as stability, switching-transient response, and noise rejection can be accurately found in short order.

A switching-mode power supply in a closed-loop configuration comprises a dc-to-dc converter (which serves as the power stage), voltage-divider network, stable voltage reference, error amplifier, compensation network, pulse-width modulator, and driver stage (Fig. 1). In operation, a portion of the

---

Reprinted with permission from *Electronic Design,* Vol. 31, No. 15, July 7, 1983; copyright Hayden Publishing Co., Inc., 1983.

Table 1. Various Networks for Compensation and Their Transfer Functions. Continued through page 63.

NETWORK	TRANSFER FUNCTION	COMMENTS
R1 series, R2 shunt	$\dfrac{R2}{R1+R2}$	Constant Gain, Zero Phase Shift
R1 series, C1 shunt	$\dfrac{1}{1+SR1C1}$	One Pole At $f_P = \dfrac{1}{2\pi R1C1}$
C1 series, R1 shunt	$\dfrac{SR1C1}{1+SR1C1}$	One Zero At Zero Frequency One Pole At $f_P = \dfrac{1}{2\pi R1C1}$
R1 series, C2∥R2 shunt	$\left(\dfrac{R2}{R1+R2}\right)\dfrac{1}{1+SC2(R1\|R2)}$	A Pole At $f_P = \dfrac{1}{2\pi C2(R1\|R2)}$
R1 series, R2+C2 shunt	$\dfrac{1+SR2C2}{1+SC2(R1+R2)}$	One Pole At f_P And One Zero At f_z $f_P = \dfrac{1}{2\pi C2(R1+R2)}$ $f_z = \dfrac{1}{2\pi R2C2}$
C1-R1 series, R2 shunt	$\dfrac{SR2C1}{1+SC1(R1+R2)}$	One Zero At Zero Frequency One Pole At $f_P = \dfrac{1}{2\pi C1(R1+R2)}$
C1-R1 series, R2+C2 shunt	$\left(\dfrac{C1}{C1+C2}\right)\dfrac{1+SR2C2}{1+S(R1+R2)(C1\|C2)}$	One zero At $f_z = \dfrac{1}{2\pi R2C2}$ One Pole At $f_P = \dfrac{1}{2\pi (R1+R2)(C1\|C2)}$ (OR)

61

NETWORK	TRANSFER FUNCTION	COMMENTS
R1, C1, R2 passive network	$\left(\dfrac{R2}{R1+R2}\right)\dfrac{1+SC1R1}{1+SC1(R1\|R2)}$	One Zero At $f_z = \dfrac{1}{2\pi R1C1}$ One Pole At $f_P = \dfrac{1}{2\pi C1(R1\|R2)}$
R1, C1, C2, R2 passive network	$\left(\dfrac{R2}{R1+R2}\right)\dfrac{1+SC1R1}{1+S(C1+C2)(R1\|R2)}$	One Zero At $f_z = \dfrac{1}{2\pi R1C1}$ One Pole At $f_P = \dfrac{1}{2\pi(C1+C2)(R1\|R2)}$ (OR)
R1, C1, R2, C2 passive network	$\dfrac{SR2C1}{1+S(R1C1+R2C2+R2C1)+S^2R1R2C1C2}$	One Zero At Zero Frequency Two Poles At $f_P = \dfrac{1}{2\pi\sqrt{R1R2C1C2}}$
R1, R2 op-amp	$\dfrac{R2}{R1}$	Constant Gain Zero Phase Shift
R1, C2 op-amp	$\dfrac{1}{SR1C2}$	A Pole At Zero Frequency
R1, R2, C2 op-amp	$\dfrac{1+SR2C2}{SC2R1}$	One Pole At Zero Frequency One Zero At $f_z = \dfrac{1}{2\pi R2C2}$
R1, R2, C2 op-amp	$\left(\dfrac{R2}{R1}\right)\dfrac{1}{1+SR2C2}$	A Pole At $f_P = \dfrac{1}{2\pi R2C2}$

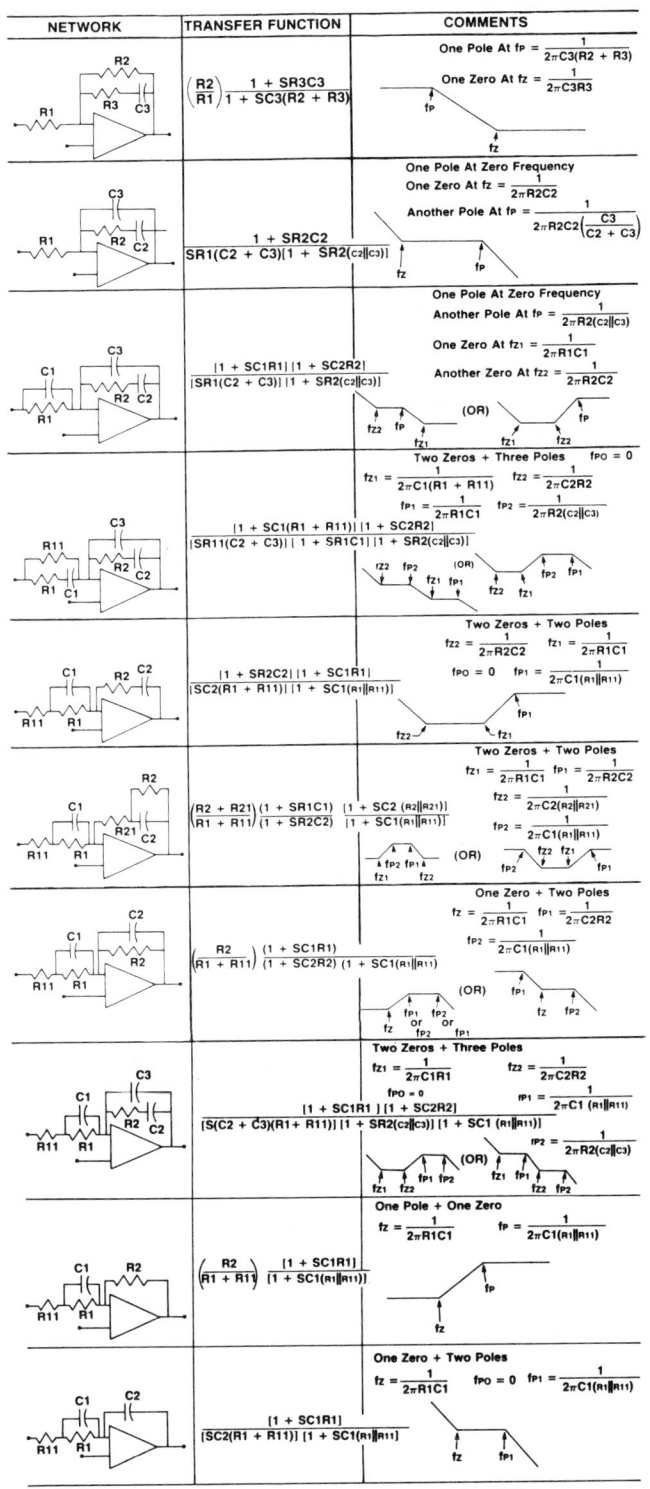

NOTE: PHASE INVERSION (180°) DUE TO OP AMP IS NOT INCLUDED IN 11 TO 24

Modeling a Switching-Mode Power Supply

Switching-mode regulators operate by storing energy in an inductor during one half of a power cycle and then transferring it to a capacitor in the following half. In contrast, dissipative-series regulators drop the difference in voltage between the input and output across a variable resistor (a transistor in its linear or conduction region) to keep the output constant. Because the transistor in a switched-mode regulator is operated either in saturation or cutoff, and because its inductors and capacitors ideally have no loss, switching regulators are highly efficient.

Modeling aids the designer in setting the supply's loop gain and bandwidth and allows the determination of its transient response and line-noise rejection.

The first step in design is to develop small-signal, low-frequency equivalent circuits for each of the blocks and then to determine their individual transfer functions. For circuits such as pulse-width modulators, error amplifiers, and voltage-divider networks, the transfer function may be written virtually from inspection (see the figure). The transfer functions for the basic dc-to-dc converters are generally known. (If unknown, however, they can be derived by performing a four-terminal measurement on the devices.)

The major task then left for a designer is to develop a suitable compensating network to ensure that the entire system will be stable. Once that is done, the exact loop gain is then determined by multiplying the gains of all the building blocks. Similarly, the phase shift is the sum of all the phase shifts of all the blocks. The switcher's basic properties—which include dc regulation, transient response, and noise rejection—can be found from the final gain equation. They can also be determined from inspection of the gain vs frequency plot or gain-margin vs phase-margin curve shown in the figure.

Although it is a small part of the circuit, the compensating network is the key to achieving system stability. In designing such a network, assume the gain of the error amplifier to be unity and assume initially it does not contain any frequency-dependent terms. The dc loop gain of the system is given by:

$$T = \left[\frac{V_{out}(s)}{d(s)}\right]\left(\frac{R_A}{R_A + R_B}\right)\left(\frac{1}{V_M}\right)A(s)$$

The response of the power stage used in a con-

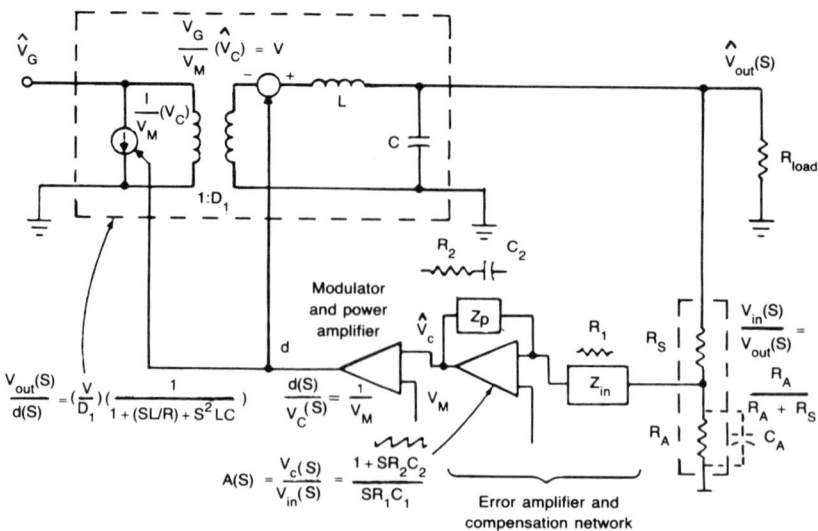

ventional supply provides a power stage slope of −40 dB from an initial frequency equal to the corner frequency of the power stage's filter. Note that the value of the load resistance, the L and C, and the characteristic resistance of the filter determine the filter's damping factor. A 0-dB or reference line can be set to define the system's operating point on the gain vs frequency plot by properly selecting the resistor values in the dc gain term. Let x dB define the 0-dB line. A double-sided arrow across that line shows that the line can be moved up or down either by varying the dc gain term or by raising the gain of the error amplifier to the required level.

The system will oscillate if the loop is closed without any compensation because the phase shift is 180° at the 0-dB crossover point. Therefore, a pole (fp) with a −20-dB/decade rolloff is placed as shown to eliminate the problem. Again, the double-sided arrow across the −20-dB slope line indicates that the pole frequency may be moved to the left or right. If it is moved to the left, the bandwidth will be less but the system will be stable. However, the transient response will be poor. For that reason, the pole is moved to the right so that it cuts the 0-dB line at a frequency (f_{po}) that is less than the corner frequency of the power stage filter, (f_{lc}). Now the system will be stable since the loop gain crosses the 0-dB line with −20-dB slope, and the gain and phase margins can be easily determined. Moreover, the transient response of the system will be better because the bandwidth has increased. However, a further increase in bandwidth is not possible in this case because the corner frequency of the power stage's filter is in the neighborhood of frequency f_{po}. An additional improvement in the loop gain can be achieved by placing a pole-zero pair with the (f_z) frequency lower than f_p. This increases the low frequency gain, which results in improved line and load regulation. From the frequency plot, it can be seen that the dominant pole can be set practically by adding an appropriate capacitor to the voltage divider network. The feedback loop of the error amplifier must be modified accordingly. The specific loop gain of the supply is thus:

$$T = \frac{V}{D_1}\left[\frac{1}{1+(SL/R)+S^2LC}\right]$$

$$\left(\frac{1+SR_2C_2}{SR_1C_1}\right)\left(\frac{R_A}{R_A+R_B}\right)\left(\frac{1}{V_M}\right)$$

The input-to-output transfer function, which is a measurement of the system's rejection of line noise, is given by:

$$F = \frac{V_{out}(s)}{V_G(s)} = \frac{D_1}{1+(SL/R)+S^2LC}$$

In any closed-loop, the line-rejection function is F' = F/T.

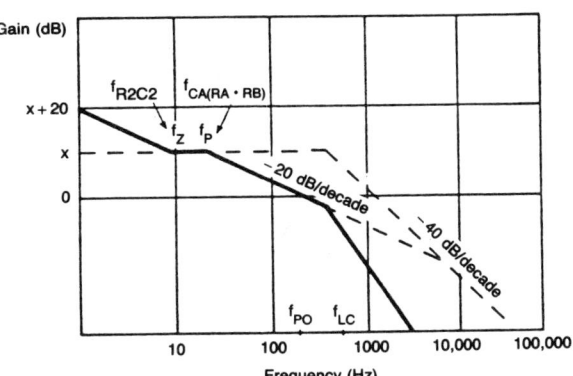

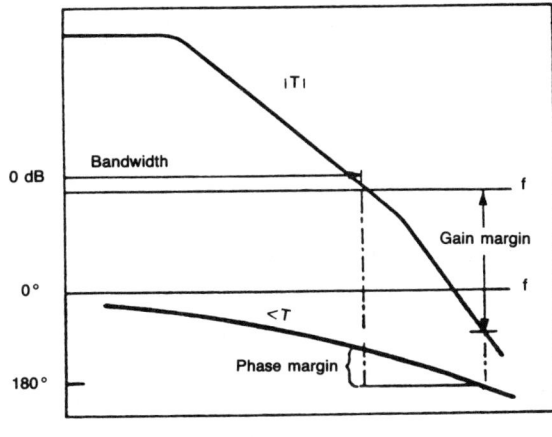

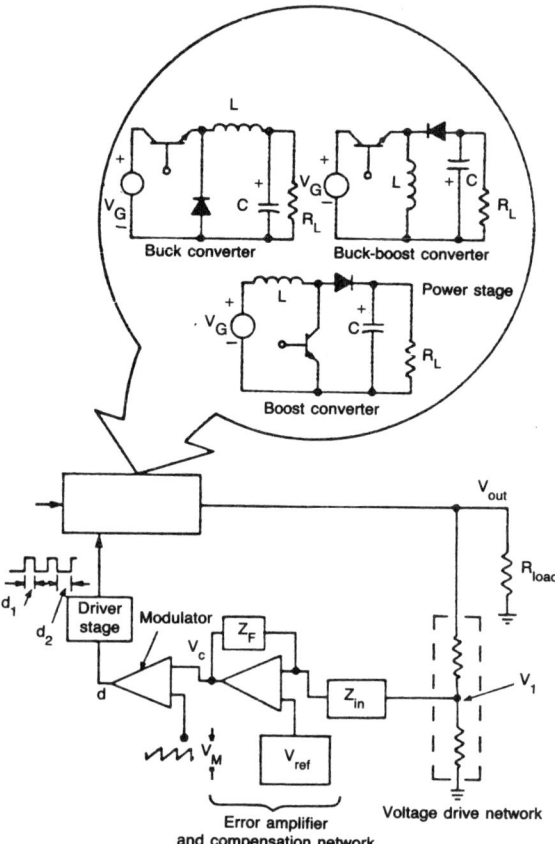

Fig. 1. A conventional Switching Mode Power Supply contains a buck, boost, or buckboost converter (Power Stage), an error Amplifier, and a pulse width modulator. A designer must use the equivalent ac circuit of each building block in order to determine the basic frequency response of the supply. Once that is done, the supply's characteristics (dc regulation, transient response, etc.) are determined.

linear circuits in the supply (see "Modeling a Switching-Mode Power Supply").

If the circuit illustrated in Fig. 1 is redrawn to show the ac equivalent of the power stage, the appropriate transfer functions that characterize the supply can be derived. It will then be apparent that all of the principal properties of a regulator, that is, dc regulation, output impedance, transient response, and line rejection, are closely related to the loop gain of the circuit. In other words, the magnitude and phase of the loop gain must be measured accurately to verify the regulator's basic characteristics. Thus measuring frequency response will, ultimately, reveal a system's stability, ability to reject input noise and ripple, and transient response.

Open-loop testing represents an inexpensive approach to making frequency-response measurements. Although it is not as accurate as closed-loop driving, it is a reasonably precise approach. One simple technique was developed by Hewlett-Packard[1] using a method often employed for testing linear feedback systems.[2] Based on a phasor-triangle technique, the method uses frequency-selective, narrow-band voltmeters to find the magnitude of the input and output voltages. Gain and phase are then determined from those readings.

output voltage is compared with a reference voltage through the error amplifier in the feedback system. The resultant error signal is amplified sufficiently to drive a modulator that delivers a PWM output voltage back to the power stage.

1.0 The Three Converter Blocks

The power stage may be configured as a buck, boost, or buck-boost converter. All other converters are derived from these three types. Basically nonlinear devices, the converters nevertheless have relatively simple equivalent circuits, as do the other

Driving an Open Loop

In an open-loop approach, a measurement is made by first opening the loop of a system with negative feedback and connecting a load impedance that simulates the closed-loop impedance at that point. A signal is then injected in the forward path, resulting in a voltage at the output. The ratio of these two voltages yields the loop gain for a particular frequency. Of course, loop gain and phase will vary as the frequency of the injected signal is swept over the range of interest.

An open-loop test jig (Fig. 2) employs a wave analyzer to generate a test signal at a given frequency in order to drive the switching regulators. The magnitude and phase of response of signal B with respect to input signal A are determined by

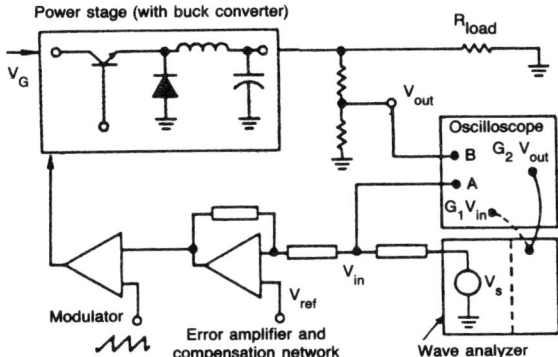

Fig. 2. Measuring the open-loop frequency response of the supply to verify its characteristics will yield accurate results if performed carefully. The open loop method, however, has inherent drawbacks. These include maintaining the operating point at a high gain level as tests are being carried out and the need to manually connect and disconnect the test probes.

an oscilloscope. These two signals, A and B (i.e., V_{in} and V_{out}), are processed through a dual-channel preamplifier in the scope, which provides some gain control. The preamplifier's response, which is available at the scope's vertical output, is then applied to the analyzer's voltmeter. Using the instrument's selector and inverter switches, the magnitudes of signals A, B, A + B, or A − B can be measured.

Network Analyzers Step In

For greater accuracy of phase measurements, the preamplifier gains for signals A and B should be individually adjusted so that the magnitudes of both signals will be equal. Thus measurement of phase angles will be more accurate. It is absolutely essential to use a narrow-band voltmeter so that the low-amplitude test signal will not be completely swamped by the switching noise. Although such meters are available, it is better to employ an instrument, say, a wave analyzer, that combines an oscillator and narrow-band voltmeter in such a way that the filter automatically tracks the frequency of the oscillator.

A sweep oscillator with leveled output, combined with a narrow-band frequency-selective voltmeter capable of locking on to the oscillator, can also be used to measure V_{in} and V_{out} in dB. That measurement will give not only V_{out}/V_{in}, but also the phase of V_{out} with respect to V_{in}. Voltmeter-oscillators are contained in instruments like the HP3040A and the BAFC0 916H frequency-response analyzer. These instruments, however, can be used to greater advantage in closed-loop measurement systems.

As mentioned, a narrow-band tracking voltmeter is required because a large amount of switching noise is produced in the supply. Further, its output contains high-order harmonics of the switching frequency, plus various modulation components caused by the mixing of the signal frequency with the noise. Note that a frequency selective voltmeter is required despite the fact that the power stage's output filter (corner frequency typically one-tenth or one-twentieth the switching frequency) attenuates most harmonics. A stand-alone voltmeter simply cannot distinguish between the components of the injected frequency and the switching frequencies.

In addition to these difficulties, the loop has to be maintained at the same operating point at all times. In addition, it is very difficult, if not impossible, to test high-gain systems in the open loop as they either saturate or cut off. Also, matching or determining the closed-loop impedance (which has to be connected when the loop is opened) is difficult. And if the system has large bandwidths, say, in the neighborhood of half of the switching frequency, there can be a difference in the measurements between open loop and closed loop because of large ripple voltages.

Magnetic Injections

Closed-loop testing, which has none of these drawbacks, is obviously the preferred approach. Moreover, because a loop does not have to be opened, measurements can be made in a very small amount of time. As noted, up until now the closed-loop approach required special equipment. But a method known as voltage injection uses a readily available current probe to insert a signal into the feedback loop, introducing it at a point where the signal is confined to a single path. What's more, the current probe does not alter the feedback loop since it has a very low output impedance. In an alternative approach, called current injection, a floating ac voltage source (whose frequency can be swept over a range) can be connected at any point in the feedback loop where the output impedance is much less than the input impedance.

In either case, the equivalent output impedance of the device that the probe is connected to should be much smaller than the input impedance at that point of injection. Both signal injecting techniques are even applicable to switching regulators with multiple feedback loops because the feedback signal from the modulator, with duty ratio, d (see Fig. 1, again), is the only control input to the power stage. Moreover, all the feedback signals are summed up at one input to drive the pulse-width modulator. The summing junction usually satisfies the impedance requirements and hence is an ideal place to inject a signal. In most switching regulators, the point that usually meets these criteria is either immediately following the error amplifier (in series with the control voltage to the modulator) or immediately following the output filter capacitor (in series with the input to the error amplifier).

Once those criteria have been met, frequency-response measurements can be taken. Let the injection voltage (signal) be V_s, which results in a signal voltage, I_{in}, at the input (forward path). V_{in} passes through the power stage and the error amplifier and results in a voltage of V_{out}. These three voltages, V_{in}, V_{out}, and V_s form a vector triangle. The loop gain is then V_{out}/V_{in}, provided that the input and output impedance criteria are met. At low frequencies, the loop gain is high and V_{out} is much higher than V_{in}. At high frequencies, the loop gain is normally low, and V_{out} is much smaller than V_{in}. From the vector triangle, the angle between V_{in} and V_{out} can easily be determined by using the well-known geometrical relationship $a^2 + b^2 - 2ab(\cos \theta) = c$, where $a = V_{in}$, $b = V_s$, and $c = V_{out}$. As the values of all the sides are already known, the phase can be determined without any trouble.

Tempus Fugit

Closed-loop measurements (Fig. 3) are more accurate because this approach takes advantage of the automatic swept-frequency capabilities of the test equipment. Further, measurements with the open-loop jig usually take several days because it is a manually set system. However, the closed-loop set up needs only a few minutes to perform the test. In the closed-loop approach, the gain of the loop is found, directly in decibels, by subtracting the input signal, A, from the output response, B. The equipment can also display B – A, if desired, and can measure the phase difference between the two inputs. Additionally, the output of the phase-detecting circuit can be connected to an X-Y-Z plotter to allow magnitude vs frequency and phase vs frequency plots to be obtained very quickly.

The two important transfer functions to be measured are $H(s) = V_G/V_{out}$ and d/V_{out}. The former yields information on the ability of the circuit to reject input noise or ripple. The latter, which is the duty-cycle-to-output response, yields information on variations in the output as a function of variations in the width of an input pulse. (In a

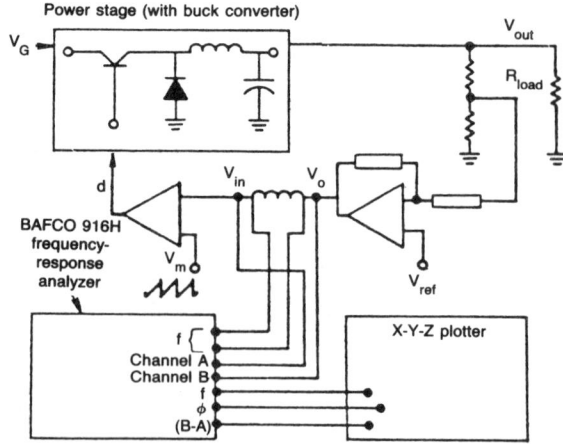

Fig. 3. Closed loop measurements of a Switching Mode Supply do not face the difficulties associated with openloop driving. Moreover, because the loop remains closed, the test equipment can run checks quickly and automatically. Two major transfer functions, the input to output response and the loop response, which measure input-noise rejection, and stability and transient response, respectively, can be easily measured with the setup here.

closed-loop system, the duty ratio is determined by the compensated and amplified error-feedback signal.) This system can measure both transfer functions. In order to determine the first, the duty cycle must be held constant. The second can be found by holding the input supply voltage (V_G) constant.

References

1. "A Quick, Convenient Method for Measuring Loop Gain," *Hewlett-Packard Journal*, Vol. 14, January-February 1963, pp. 5-8.
2. R. D. Middlebrook, "Improved Accuracy Phase Accuracy Measurement," *International Journal of Electronics,* Vol. 40, No. 1, January, 1976, pp. 1-4.

MEASUREMENT OF MAGNITUDE AND PHASE OF SWITCHING REGULATOR TRANSFER FUNCTIONS AND LOOP GAIN

Techniques to measure the magnitude and phase of switching regulator functions and loop gain are discussed. Besides, an introduction to power electronics and a review of modeling is presented.

Introduction

In the 1960s, demands of the space programs led to the development of highly reliable, efficient and lightweight electrical power systems for spacecrafts. Despite limited supply of available energy onboard the spacecrafts, engineers found innovative solutions for power processing and management of the electrical power onboard the spacecraft. These helped usher in the era of modern power electronics. Today, similar limitations on sources of available energy are becoming a prime design consideration in everyday electric power processing.

Power electronics is entirely devoted to switched mode power conversion and deals with modern problems in analysis, design and synthesis of electronic circuits as applied to efficient conversion, control, and regulation of electrical energy. Design and optimization of dc-to-dc converters, which offers the highest power efficiency, small size and weight, and high performance are also included in power electronics.

These dc-to-dc converters with isolation transformers can have multiple outputs of various magnitudes and polarities. Regulated power supply of this type has wide applications, particularly in computer systems, wherein low voltage high current power supply with low output ripple and fast transient response are mandatory. In addition, these converters connected in a particular configuration result in switched mode ac power amplifiers with enough bandwidth and high efficiency. Off-line switches, dc-to-dc single and multiple output power supplies, bi-directional power supplies (battery chargers and dischargers), dc-to-ac inverters, dc-to-ac uninterruptible power supplies, dc-to-ac motor control, power servo control, robotics, and switching audio amplifiers etc. are some of the examples of switch mode power systems.

Switching-mode power supplies have come into widespread use in the last decade. An essential feature of efficient electronic power processing is the use of semiconductors in a power switching mode (to achieve low losses) to control the transfer of energy from source to load through use of pulse width modulated control techniques. Inductive and capacitive energy storage elements are used to smooth the flow of energy while keeping losses to a low level. As the frequency of switching increases, the size of the magnetic and capacitive elements decreases almost in direct proportion. Because of their superior performance, i.e., high efficiency, small size and weight and relatively low cost, they are displacing conventional linear (dissipative as they operate in linear or conduction mode) power supplies even at power levels as low as 25W.

The modeling, analysis and design of these switching dc-to-dc converters have been extensively

---

Reprinted with permission from *CSIO Communications Journal,* India, Jan.-March 1982.

carried out in the past five years and it is commonly believed that the designs in commercial use today employ the simplest possible converter topology for dc-to-dc conversion. Industry has been quick to realize that the energy saving technique also affords the opportunity to make dramatic reductions in equipment size and weight. Consumer and industrial applications are expanding rapidly.

Among the various approaches developed for modeling and analysis of the switching dc-to-dc converters, the current injected equivalent circuit approach[1] and state space average approach[2] are used in producing a linear equivalent circuit model which correctly represents the non-linear converter properties for the static as well as dynamic ac small signal at low frequency levels, the essential features of the input and output transfer properties. Availability of the above model allows choice of the best converter for a specific application and optimization of the feedback loop of a regulator containing such a converter.

Also, the models enable to design the switching regulators for stable operation with large bandwidth, fast transient response and good line rejection. This is because now the design can apply standard method or circuit analysis applicable to linear feedback control systems using linear feedback theory. However, the validity of these models and the design can be made by measuring the frequency response of the system to check the accuracy of the loop gain and phase. Thus, in the design of any feedback system it is necessary to make measurements of the loop gain on the practical circuit as a function of frequency to ensure that the circuit operates as analytically predicted or get feedback from the measurement to correct the analytical prediction.

Very recently, a simple but accurate measurement method described in Hewlett-Packard Journal[3] and application note[4] for linear feedback systems was extended and successfully applied to switching regulators[5&6]. This is based on a phasor triangle technique. In this method, only magnitude measurements using voltmeters are made and gain and phase are determined from these magnitude measurements. Extension of this method to switching regulators is made by the use of frequency selective tracking narrow band voltmeters instead of simple voltmeter. Such measurements were first carried out by opening the feedback loop of a system. Later, similar measurements were carried out without opening the loop because of various associated advantages.

This paper mainly discusses the techniques and the need to measure the magnitude and phase of switching regulator transfer functions and loop gain. An introduction to modern power electronics is followed by a review of modeling of switching converter-regulators. The measurement methods including point of signal injection, open loop and closed loop measurements, equipment used are also presented.

Modeling of Switching dc-to-dc-Converters

In the last five years, modeling and analysis of switching dc-to-dc converters has received considerable attention because of high performance requirements of power processing systems and the efforts in this direction have resulted in the characterization of transfer as well as input and output properties of basically non-linear switching dc-to-dc converters in the frequency domain. Among the various approaches attempted to attain this, the current injected equivalent circuit approach and state space average approach are worth mentioning. The current injected equivalent circuit approach is briefly reviewed below:

The following conventions and notations are followed in the modeling and analysis:

$d1Ts + d2Ts = Ts$ and $Ts = 1/f_s =$ Switching period where

$d1Ts =$ the interval during which the transistor is turned on and the diode is off and

$d2Ts =$ the interval during which the transistor is turned off and the diode is on.

Besides, the capitalized quantities are used for steadystate values and the quantities with dots for

1. SWITCHING DC-DC CONVERTERS

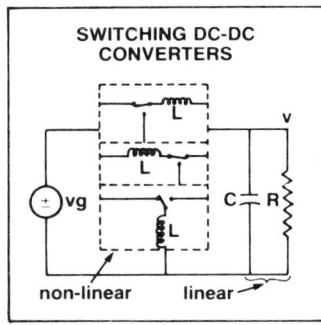

non-linear | linear

2. CONVERTER EQUATIONS

i) Derivative of inductor current

ii) Average inductor current in a switching period (i_{ave})

iii) Output voltage $(v) = i_{ave} \cdot z$ (z = impedance of output network)

3. STEADYSTATE PROPERTIES

Derivative of inductor current = 0
$vg \to Vg$
$v \to V$
$z \to Re(z)$
$i_{ave} \to I_{ave}$
$d1 \to D1$
$d2 \to D2$

4. PERTURBATION & LINEARIZATION

$d1 = D1 + \hat{d1}$
$d2 = D2 + \hat{d2}$
$vg = Vg + \hat{vg}$
$v = V + \hat{v}$
$i_{ave} = I_{ave} + \hat{i}_{ave}$

Perturbation product terms neglected to obtain once again linear system

5. DYNAMIC PROPERTIES

i) Input to output transfer function

ii) Control to output transfer function

6. LINEAR EQUIVALENT CIRCUIT (Boost)

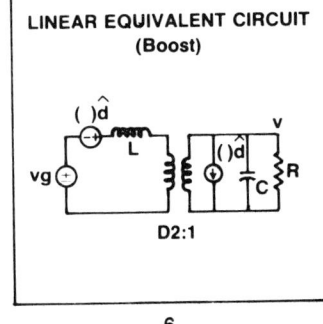

Fig. 1. Flowchart of current injected equivalent circuit approach to modeling switching dc-dc converters in CIC mode.

the small perturbations.

The current injected equivalent circuit approach to modeling converters operating in continuous inductor conduction mode is outlined in flowchart (Fig. 1), which is very general and applicable to various power stages. The first step in this process is to identify non-linear and linear parts of the converter circuit and linearize the non-linear part of the converter (Box 1). The non-linear part of the converter determines the average current injected into the linear part. A set of relationships are written (Box 2) referring to the converter diagram and current and voltage waveforms shown in Fig. 2.

The steady state solution is achieved by setting derivatives and perturbations to zero (Box 3). Since the converter equations in Box 2 are linear, superposition holds and can be perturbed (Box 4) by the introduction of a small ac variation over the steady state operating point. As the independent driving

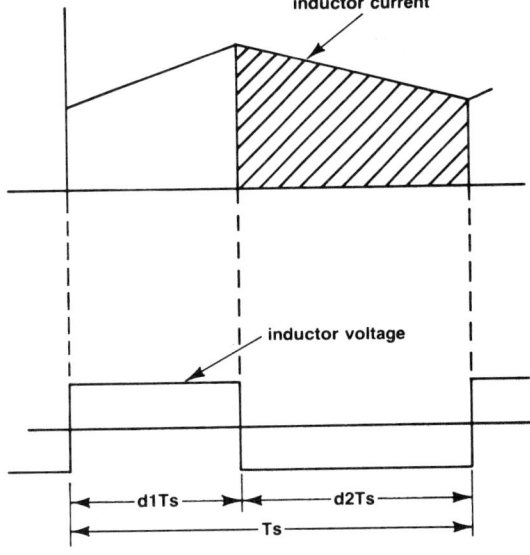

Fig. 2. Typical inductor current and voltage waveforms.

inputs are vg and d, the perturbation in these two inputs cause the perturbation i and v. Now, make the small signal approximation, namely, the small ac variation from the steady state operating point values, i.e., $\hat{v}/V$, $\hat{v}g/Vg$, $\hat{d}1/D1$, $d2/D2$, $\hat{i}/I$ (each) < < 1. Using the above approximation, non-linear second order terms are neglected to obtain once again linear set of equations. Thus, only the ac part is retained which describes the small signal low frequency behavior of the converter. Using these set of equations, the input to output and control to output transfer functions (Box 5) are written. Using the same set of equations, an equivalent circuit (Box 6) is drawn which represents the input and output small signal low frequency properties of the non-linear converter.

Modeling of Buck, Boost and Buckboost Converters

The current injected equivalent circuit approach to modeling converters reviewed above is applied to buck, boost and buckboost converters and the important results i.e., input to output and duty ratio to output transfer functions are presented along with their small signal low frequency equivalent circuits (Figs. 3, 4 and 5).

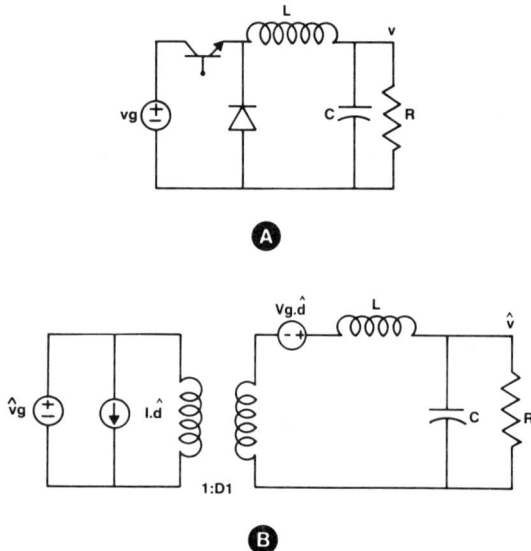

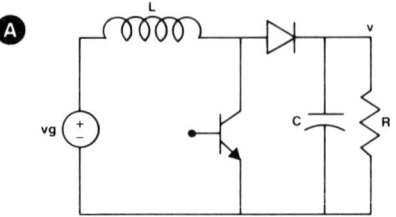

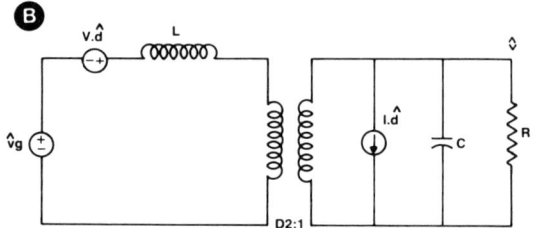

Fig. 4. (A) Boost converter. (B) Its small signal low frequency linear equivalent circuit model.

Buck Converter

$$\frac{\hat{v}(s)}{\hat{v}g(s)} = (D1) \frac{1}{1 + \frac{SL}{R} + S^2LC}$$

$$\frac{\hat{v}(s)}{\hat{d}1(s)} = \left(\frac{V}{D}\right) \frac{1}{1 + \frac{SL}{R} + S^2LC}$$

Boost Converter

$$\frac{\hat{v}(s)}{\hat{v}g(s)} = \left(\frac{1}{D2}\right) \frac{1}{1 + \frac{SL}{R(D2)^2} + \frac{S^2LC}{(D2)^2}}$$

$$\frac{\hat{v}(s)}{\hat{d}1(s)} = \left(\frac{V}{D2}\right) \frac{1 - \frac{SL}{R(D2)^2}}{1 + \frac{SL}{R(D2)^2} + \frac{S^2LC}{(D2)^2}}$$

Buckboost Converter

$$\frac{\hat{v}(s)}{\hat{v}g(s)} = \left(\frac{D1}{D2}\right) \frac{1}{1 + \frac{SL}{R(D2)^2} + \frac{S^2LC}{(D2)^2}}$$

Fig. 3. (A) Buck converter. (B) Its small signal low frequency linear equivalent circuit model.

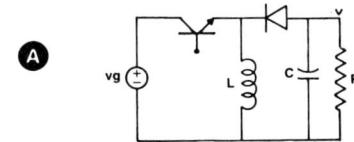

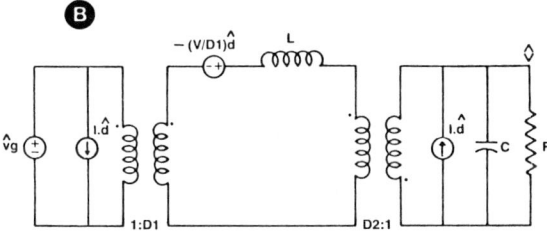

Fig. 5. (A) Buckboost converter. (B) Its small signal low frequency linear equivalent circuit model.

$$\frac{\hat{v}(s)}{\hat{d}(s)} = \left(\frac{V}{D1 \cdot D2}\right) \frac{1 - \dfrac{SL}{R(D2)^2}}{1 + \dfrac{SL}{R(D2)^2} + \dfrac{S^2 LC}{(D2)^2}}$$

The output of switching dc-to-dc converter is regulated by closing the loop with proper compensation as shown in Fig. 6. The transfer functions for various building blocks of switching regulator[7] are given below:

For Voltage Divider Network

$$\frac{\hat{v}\, l(s)}{\hat{v}(s)} = \frac{R1}{R1 + R2} = K \ (say)$$

For Error Amplifier

$$\frac{\hat{v}\, c(s)}{\hat{v}\, 1(s)} = \frac{Zf}{Zin} = A(s) \ (say)$$

For Modulator

$$\frac{\hat{d}\, 1(s)}{\hat{v}\, c(s)} = \left(\frac{1}{Vm}\right) = Gm(s) \ (say)$$

Now, the loop gain is given by:

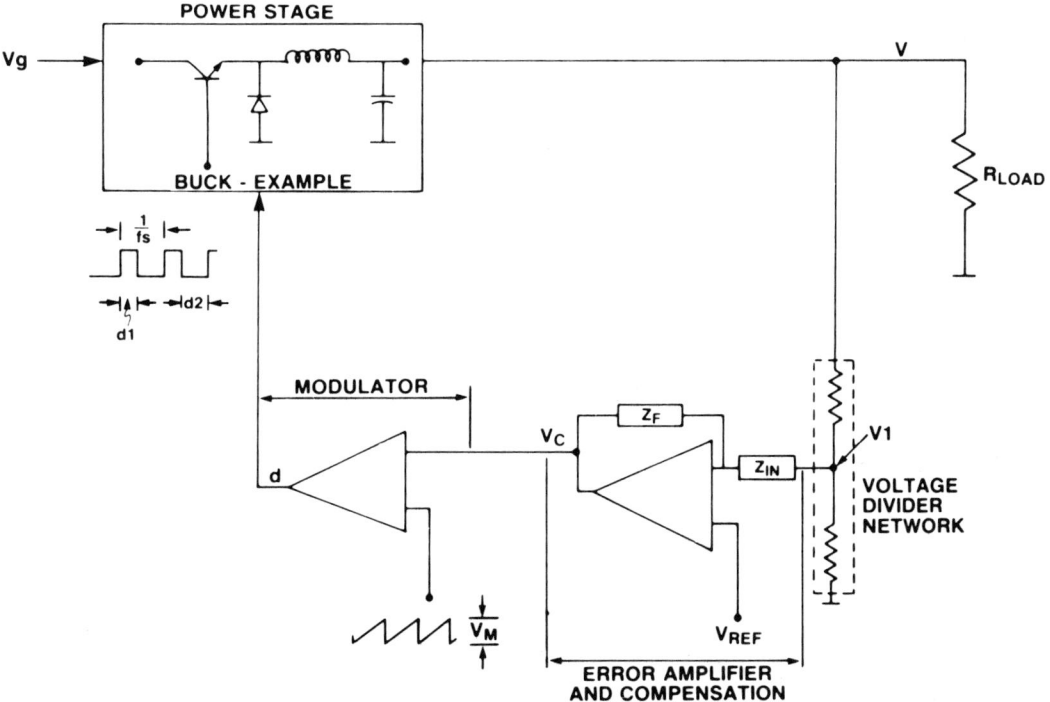

Fig. 6. Typical Switching Regulator showing its building blocks.

$$T = \frac{\hat{v}(s)}{\hat{d}_1(s)} KA(s) \, Gm(s)$$

For the converter shown in Fig. 6, loop gain is given by:

$$T = \left(\frac{V}{D}\right)\left(\frac{1}{1 + \frac{SL}{R} + S^2LC}\right)\left(\frac{R1}{R1+R2}\right)\frac{1}{Vm} A(s)$$

Loop gain is of prime importance in the analysis and measurement phases of the design of feedback circuits. Analytically, the loop gain is important because the complex frequencies, at which the loop gain attains value of one are the closed loop poles of the system. Consequently Nyquest and Bode techniques[8] can be applied to obtain stability information. In addition, measurement of the loop gain provides an excellent tool for the design verification. Thus measurements are indispensable in the design and only after comparing the measurements with theoretical predictions, the design can be treated successful and complete. This provides the motivation for frequency response measurements.

Frequency Response Measurements

Frequency response measurements reveal the stability of a system, its ability to reject the input noise/ripple propagating to the output and its transient response. Frequency measurements on switching dc-to-dc converters and regulators require special equipment as these systems are inherently non-linear and produce noise in the switching process.

Necessity for Narrow Band Tracking Voltmeter

Large amount of switching noise is produced in switching process. Even a small signal perturbation of control signal (duty ratio) at single frequency FM generates an output which contains the following frequency components;

fm, 2fm, 3fm . . . . ; fs, 2fs, 3fs . . . . ; and fs – fm, fs + fm etc.

However, the basic requirement of low ripple and use of lowpass filter whose corner frequency is practically in the range of fs/20 to fs/10, attenuates most of the harmonics. Even then a narrow band tracking voltmeter is required to measure the signal only at the frequency of the injected signal and discard all other frequencies.

Gain as well as phase of two sinusoidal signals are determined with the use of a voltmeter that measures only magnitudes. The ratio of output to input gives the gain. The phase determination is based on the phasor representation of sine waves and the fact that the angles of a triangle can be found if the three sides are known (from simple geometrical relationship).

Open Loop Frequency Response Measurements

For stable operation of any system, negative feedback is employed. Frequency measurements are made in such negative feedback system by opening the loop at an appropriate point and by connecting a load impedance which simulates the closed loop impedance at that point. Next, a signal is injected in the forward path which results in a voltage at the other end. The ratio of these two voltages gives the loop gain and phase when the frequency of the injected signal is swept over a range of interest. This is illustrated in Fig. 7 for a switching regulator.

There are some practical problems in making the open loop measurements. Firstly, loop has to be opened and operated at the same operating point. Secondly, it is difficult to operate high gain systems in open loop as they either saturate or cutoff. Besides, matching or determining closed loop impedance, (which has to be connected when the loop is opened) is difficult. If the system is designed with sophistication to achieve large bandwidths in the neighborhood of half of the switching frequency, there can be difference in the measurements between open and closed loop because of the interaction of the ripple with the ramp. However, in most system designs one can easily achieve very good performance by closing the loop at 10 to 15% of the switching frequency

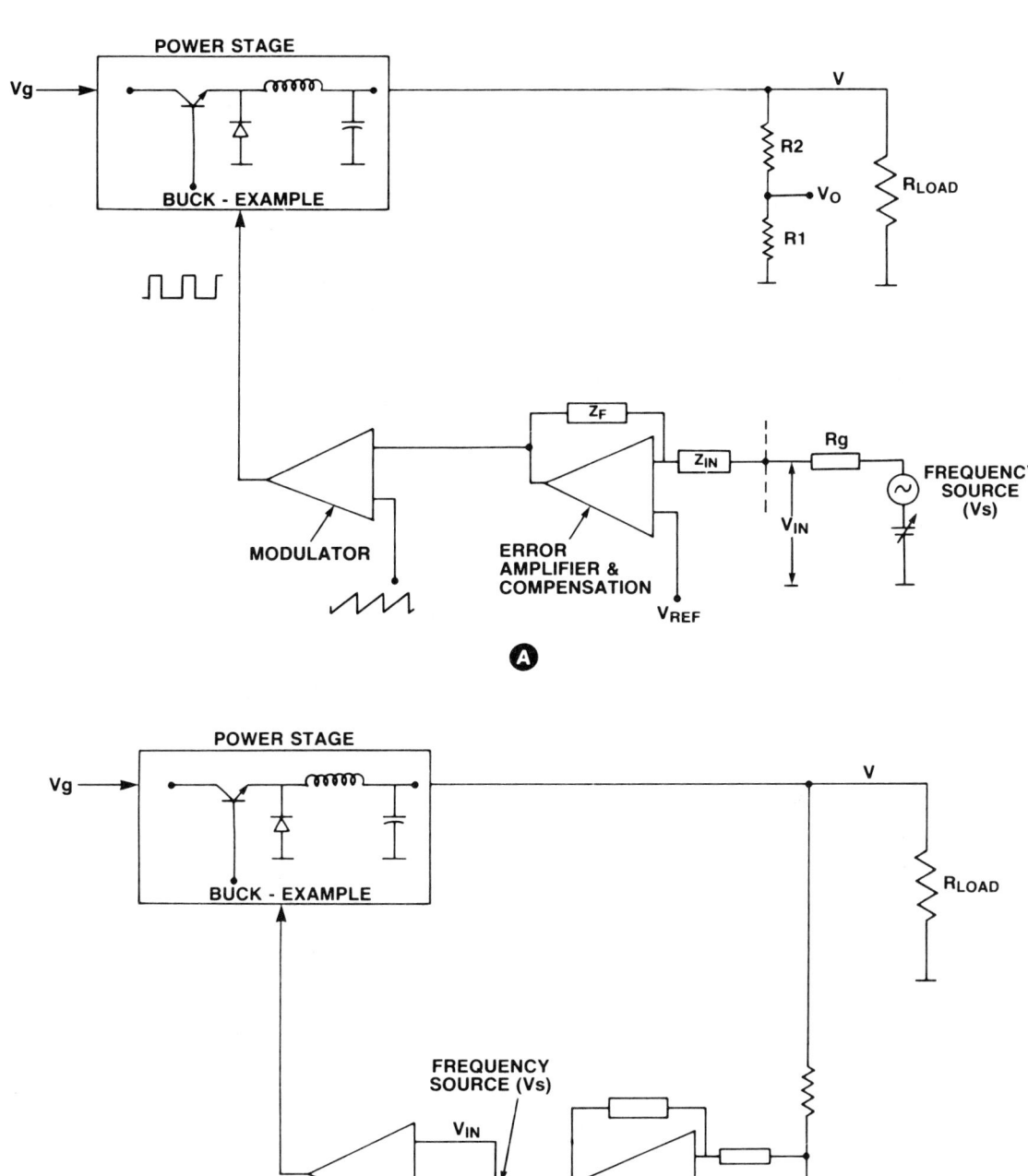

Fig. 7. Open loop frequency response measurements on a Switching Regulator (A) Approach 1 (B) Approach 2.

wherein the open and closed loop measurements results in same frequency response.

Because of these reasons, closed loop frequency response measurements are preferred. Injection of a test signal into the closed loop enables gain and phase measurements without opening the loop.

Closed Loop Frequency Response Measurements

All the demerits of open loop measurements are overcome by making the measurements without opening the loop. Consequently, the measurements can be made in a very small amount of time compared to the open loop method.

This method of making loop gain measurements utilizes the magnetic coupling capability of the current probe in association with a frequency source to insert a signal into the feedback loop (without opening it) by simply clipping it at an appropriate point (where the signal is confined to a single path) in the circuit (Fig. 8). The current probe does not alter the feedback loop since it has very low output impedance. This is known as voltage injection. Alternately, a floating ac voltage source, whose frequency can be swept over a range, can be connected at any point in the feedback loop where the output impedance is much less than the input impedance. This is known as current injection (Fig. 9).

Point of Injection Determination

If the injected signal is to be effective, neither

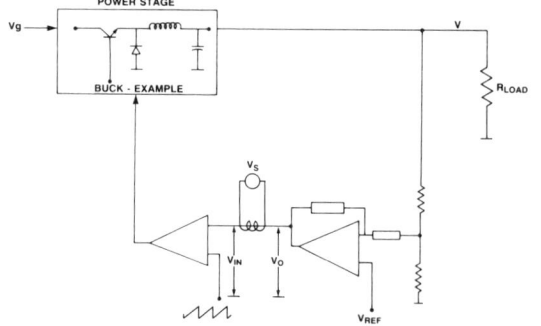

Fig. 8. Closed loop frequency response measurements using voltage injection.

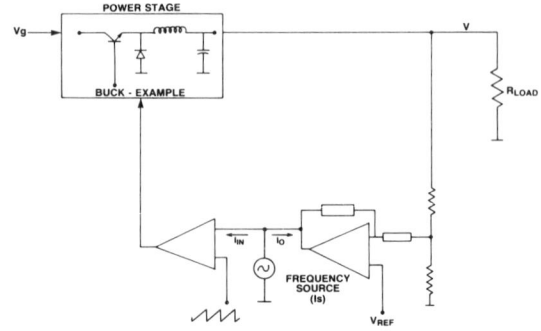

Fig. 9. Closed loop frequency response measurements using current injection.

to be attenuated nor to be interfered with the normal operation of the system, the output impedance should be much smaller than the input impedance at the point of injection (Fig. 10). This signal injecting technique at an appropriate point in the loop and making gain and phase measurements is applicable to switching regulators even with multiple feedback loops (Fig. 11). This is because duty ratio is the only control input to the power stage, all the feedback signals are summed up at one point before applying as input to the pulse width modulator. This point usually satisfies the impedance requirements and hence, is an ideal place to inject signal. In most of the switching regulators, the two points (Fig. 10) that usually meet these criteria are immediately following the error amplifier in series with the control voltage to the modulator and immediately following the output filter capacitor, in series with the input to the error amplifier.

Loop Gain and Phase Determination

Let the injection voltage (signal) be V_s resulting in a signal voltage V_{in} at the input (forward path). V_{in} passes through the power stage and the error amplifier and results in a voltage V_o as shown in Fig. 8. The loop gain is V_o/V_{in} provided the input and output impedances criteria are met. At low frequency, loop gain is high and V_o is much higher than V_{in}. The three voltages V_{in}, V_o, V_s makes a vector triangle (Fig. 12A). At high frequency, loop gain is normally low and V_o is much

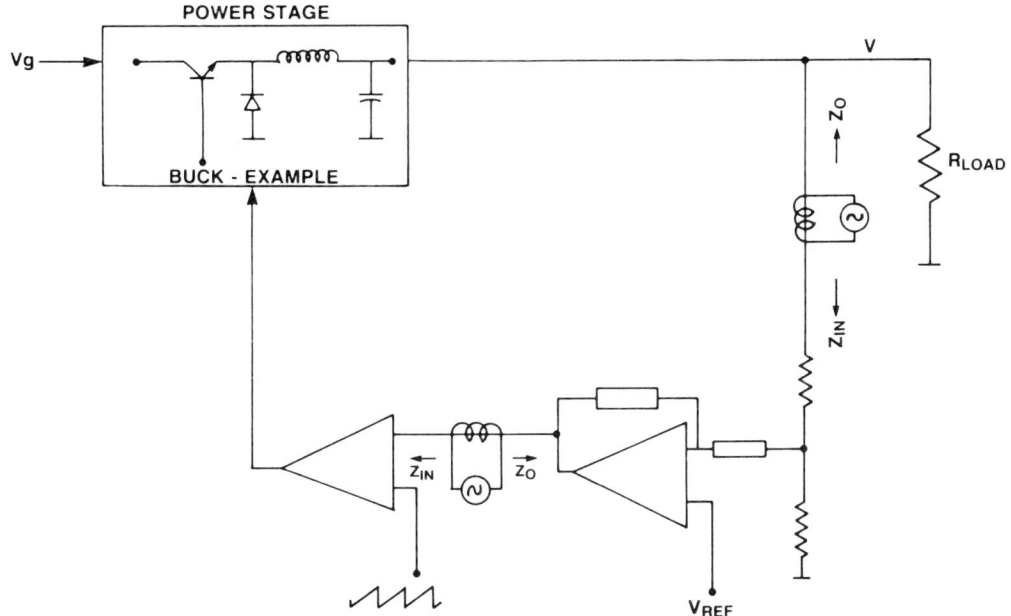

Fig. 10. Points of signal injection.

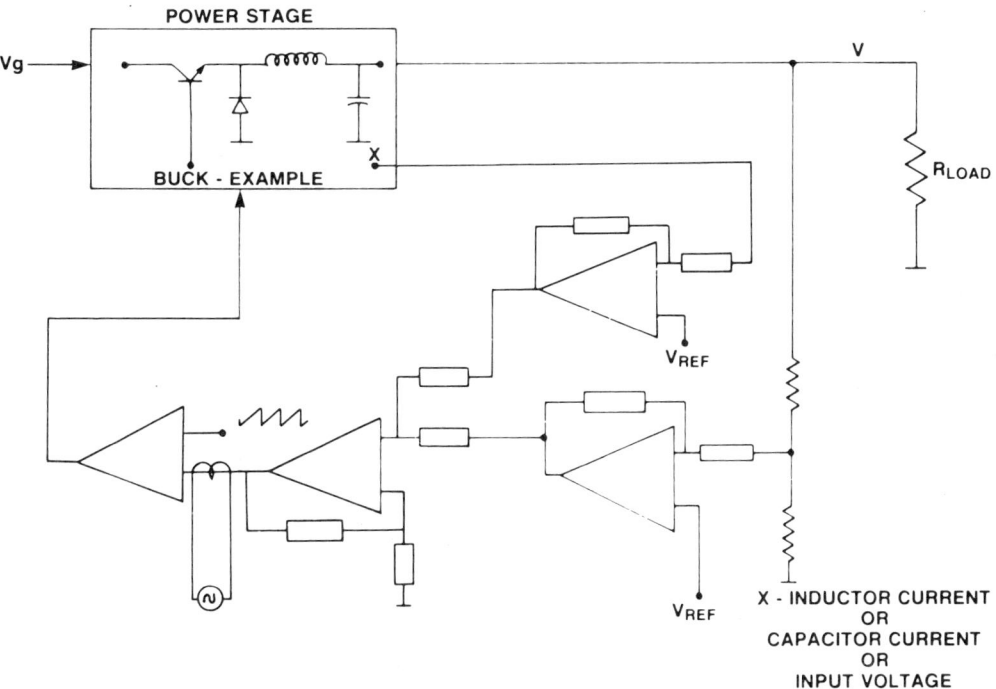

X — INDUCTOR CURRENT
OR
CAPACITOR CURRENT
OR
INPUT VOLTAGE

Fig. 11. Signal injection in a switching regulator with multiple feedback loops.

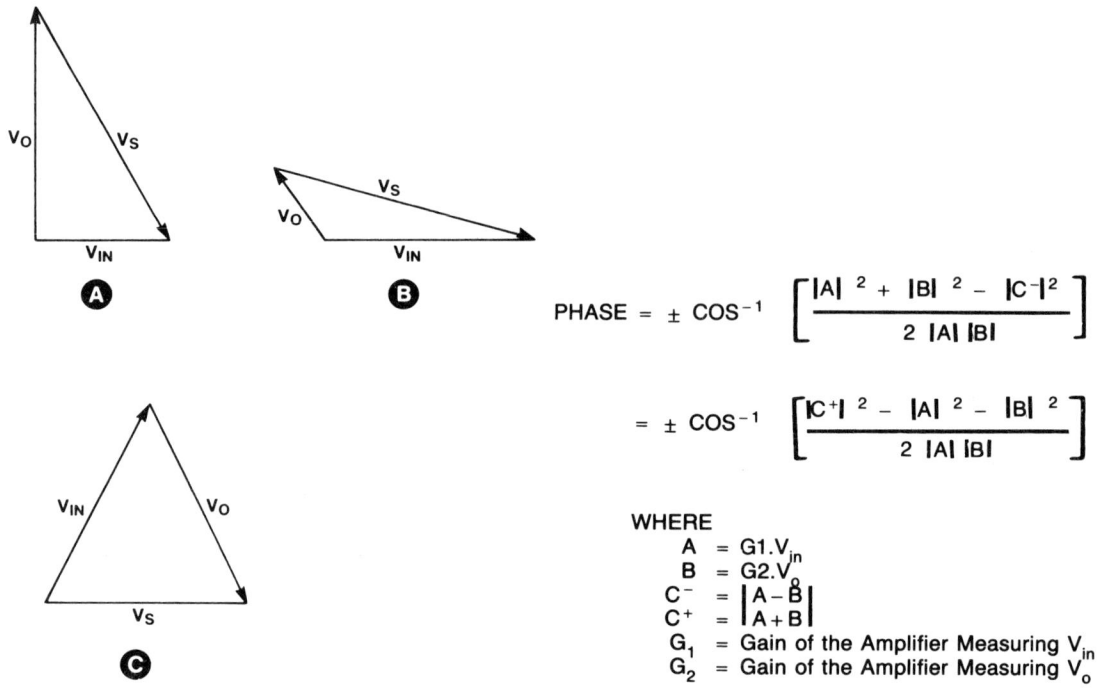

Fig. 12. Vector triangles formed by the three magnitude measurements made on a switching regulator (A) at low frequency (B) at high frequency (C) V_{in} and V_o are made equal by selecting appropriate gains for V_{in} and V_o.

smaller than V_{in}. Again, these three voltages make a vector triangle (Fig. 12B). The angle between V_{in} and V_o can easily be determined by using the geometrical relationship. As the values of all the sides are already known by measurement, the phase can be determined easily.

Measurement Set-Up 1

The measurement set-up is shown in Fig. 13 wherein a test signal at a given frequency drives the switching regulator and the magnitude and phase of signal B with respect to a signal A are to be determined.

These two signals are processed through a dual channel oscilloscope preamplifier that provides some gain control. The preamplifier output available at the oscilloscope vertical output socket is applied to a magnitude voltmeter. Using the selector and invert switches, the magnitudes of signals A, B, A+B or A−B are measured.

For better accuracy of phase measurements, the preamplifier gains for signal A and B can be different such that their magnitudes are equal. In this case, the phase triangle is an isosceles triangle and a phase measurement becomes more accurate (Fig. 12C).

The equipment needed for the above measurement are oscilloscope and wave analyzer. The oscilloscope can be Tektronix 500 series or Tektronix 7000 series. For measurements on switching regulators, it is absolutely essential to have narrow band voltmeter to avoid the small test signal being completely swamped by the switching noise. Although such voltmeters are available, it is better to use an instrument that combines an oscillator and narrow band voltmeter in such a way that the voltmeter narrow band filter automatically tracks the frequency of the oscillator. Such instruments are Hewlett-Packard wave analyzers 302A, 310A, 312A, 3590A and 3581A.

There are equipment with two channel inputs which measure V_{in} and V_o in dB and provide not only V_o/V_{in} but also phase of V_o with respect to V_{in}. Such equipment are sweeping oscillator with leveled output and narrow band frequency selective voltmeter capable of locking on to the frequency of the oscillator.

Measurement Set-Up 2

This measurement set up (Fig. 14) is more accurate compared to the earlier set-up because the equipment used here is more sophisticated and has two inputs (channels). Thus, both the signals can be measured at a time in dB. As the magnitude measurements are made in dB, subtraction of signal A from the signal B gives the gain of the loop in dB directly. This equipment has a provision to read B – A while making the measurements. It also has a phase meter which measures the phase difference between the two inputs. These outputs can be connected to a X-Y-Y plotter and frequency can be swept from f1 to f2 (range of interest) and magnitude and phase versus frequency plots can be obtained very fast. Such well known equipment are HP 3040A network analyzer and BAFCO 916H frequency response analyzer.

Measurements using the first set-up usually takes days to complete because it is manual, whereas second set-up takes few minutes and is very accurate. This is because the equipment has provision to sweep the frequency over a range and has log voltage and log frequency outputs capable of driving an X-Y-Y plotter.

Measurement of Transfer Functions

The two important transfer functions are "input to output" transfer functions which give the information about the ability of the circuit to reject input noise or ripple propagation to the output and "duty ratio to output" transfer function which gives the information about the variation in the output due to the variations in the duty ratio. In closed loop system, the duty ratio is determined by the compensated amplified error signal. These two transfer functions are measured as a function of frequency (Fig. 15).

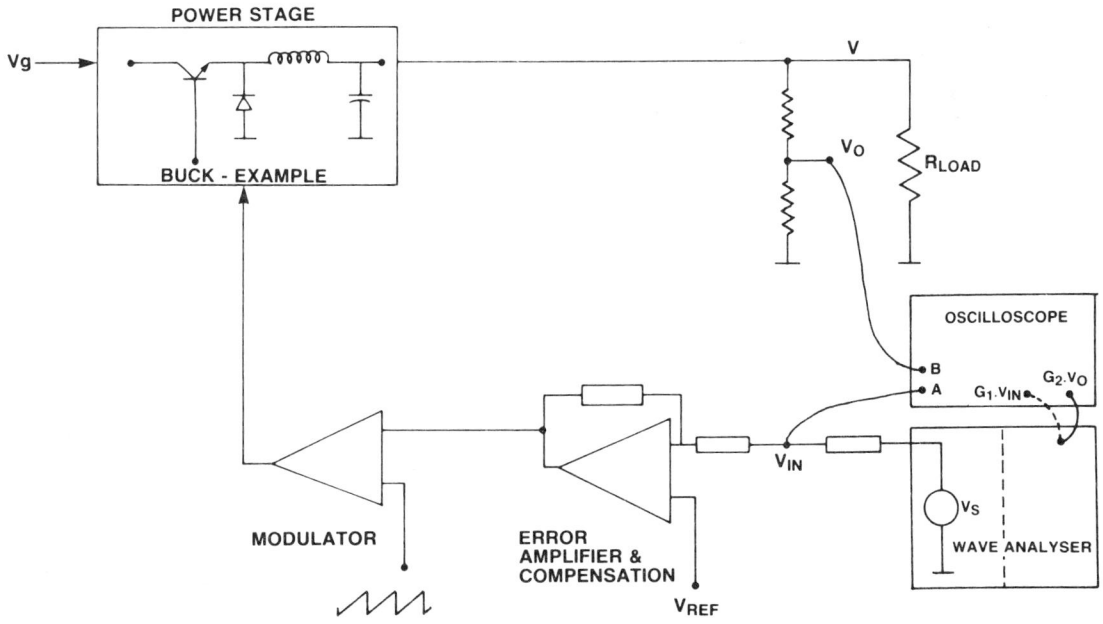

Fig. 13. Experimental set-up: open loop frequency response measurements on a switching regulator.

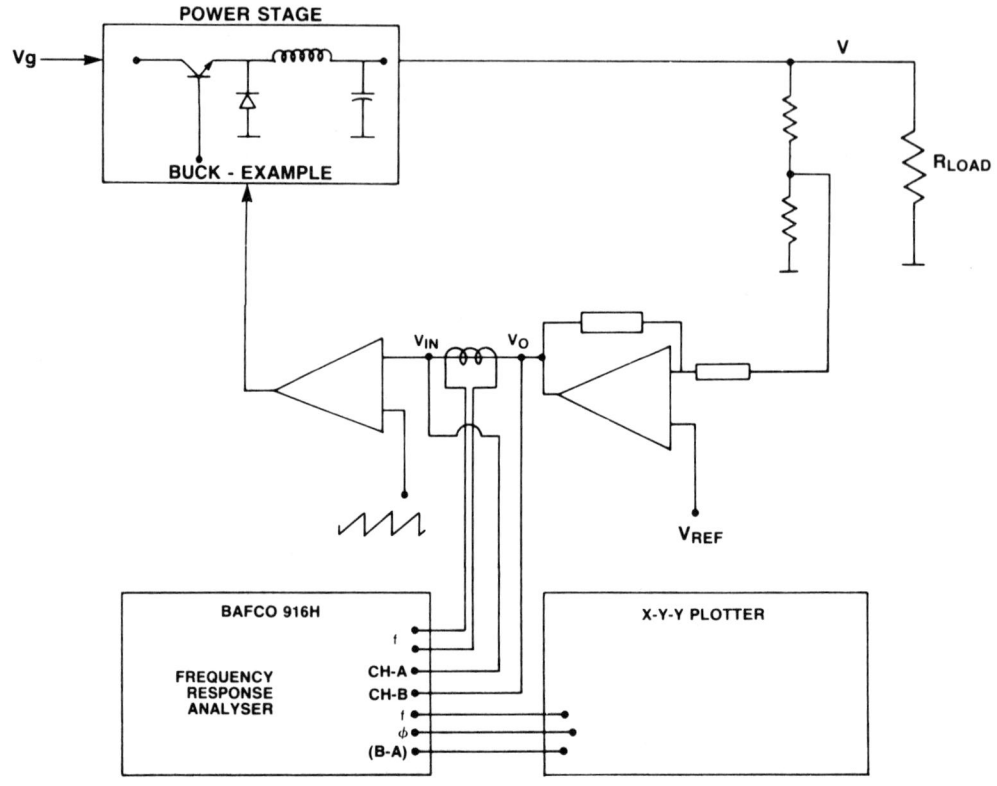

Fig. 14. Experimental set-up: closed loop frequency response measurements on a switching regulator.

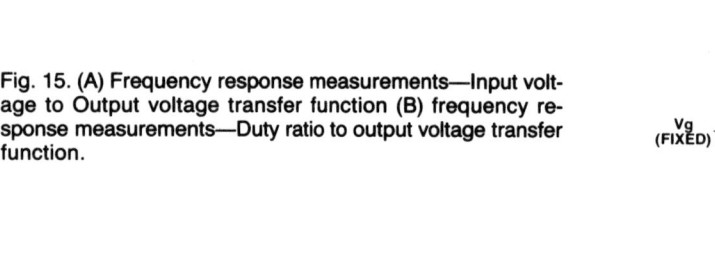

Fig. 15. (A) Frequency response measurements—Input voltage to Output voltage transfer function (B) frequency response measurements—Duty ratio to output voltage transfer function.

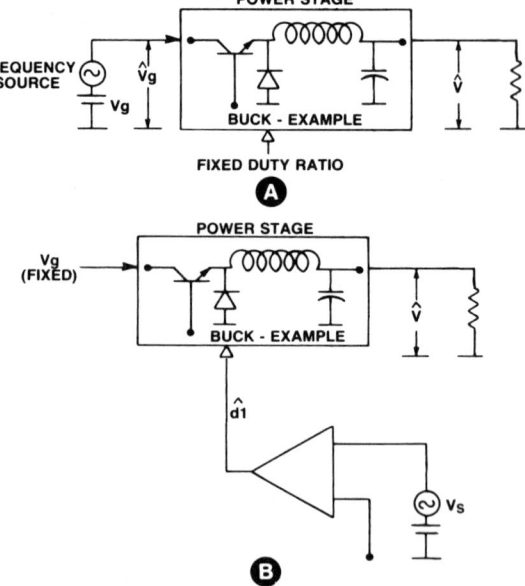

Conclusion

An introduction to power electronics and a review of modeling of switching converters and regulators are presented. The need for making measurements, extension of measurement method to switching regulators have been briefly discussed. The earlier method used to open the loop for carrying out the measurement method to switching regulators have been briefly discussed. The earlier method used to open the loop for carrying out the measurements with some practical problems. These problems are overcome by making measurements without opening the loop. Besides, loop gain and phase determination and two measurement set-ups are described.

References

1. Chetty PRK 1981 *IEEE Transactions on Aerospace Electronics and Systems* 17 (6).
2. Middlebrook RD and Cuk Slobodan 1976 *IEEE Power Electronics Specialists Conference* 18-34.
3. *Hewlett-Packard Journal 1963* January-February 5-8.
4. *Hewlett-Packard Application Note No 59* 1965 January.
5. Middlebrook RD 1975 *International Journal of Electronics* 38 (4) 485-521.
6. Middlebrook RD 1976 *International Journal of Electronics* 40 (1) 1-4.
7. Chetty PRK *Internal Report on Modeling and Design of Switching Regulators* (Sundstrand Advanced Technology Corporation, Illinois).
8. *Electronic Design* 1965 June 40-41.

Chapter 4

Computer Aided Design

SPICE-2 CAD Package for the Design of Switching Regulators   84

SPICE-2 CAD PACKAGE FOR THE DESIGN OF SWITCHING REGULATORS

Introduction

An efficient computer aided technique for the design of switching regulators described here provides designers with great advantage for avoiding the costly and time consuming trial and error laboratory practical approach. This computer aided design technique accelerates in analyzing and evaluating the performance of switching regulators. The switching regulators have gained best place in the field of power electronics and have almost replaced their dissipative linear counterparts because of the best performance advantages like small size, higher efficiency, small volume, equal reliability, etc. Though these regulators are inherently nonlinear, the modeling techniques[1] developed in the past few years resulted in successfully producing a linear equivalent circuit model valid at low frequency small signal level including input and output properties.

The important quantities that are of main concern for the design of switching regulators[2] are:

$He(s)$ = Low pass filter characteristic
Z_{ei} = Input impedance of the low pass filter
Z_{eo} = Output impedance of the low pass filter
v/v_g = Input to output transfer function
v/d = Control to output transfer function

These are open loop quantities which are modified once the loop is closed. Hence the following quantities represent the properties of principal interest in the design and analysis of the switching regulators:

T = Loop gain and Phase Vs frequency
Z_i = Closed loop input impedance of the regulator
Z_o = Closed loop output impedance of the regulator
F = Closed loop input to output transfer function

Loop gain shall have enough dc gain and shall be properly frequency shaped. Z_o determines the transient response. A lower value of Z_o and its non-peaking nature results in best transient response. On the otherhand higher value of Z_o and its peaking nature results in large over/undershoots plus it takes more time to reach steadystate. F shows its ability for closed loop regulator to prevent line voltage variations from appearing in the output. Z_i is important in determining the modified regulator properties when an input filter is added.

In the following section a switching regulator has been simulated and the above described quantities of principal interest are predicted and compared with practical measurements made in[2]. This type of computer aided design (CAD) package is very useful and handy for power supply designers. Thus, the main goal of this paper is to show the usefulness of SPICE-2 CAD package for the design of the switching regulators. Also SPICE-2 circuit analysis package offers an easy approach to simulate the switching regulators with the help of presently available modeling approaches using low frequency small signal linear equivalent circuit for inherently non-linear switching regulators.

Simulation

For verification purposes a 10 volt, 1 amp switching regulator circuit[2] as shown in Fig. 1 is simulated. This is a buck switching regulator operating at a frequency of 100 kHz and the results will be compared with the practical measurements. Figure 2 gives the equivalent circuit ready for SPICE-2 input. Figure 3 gives the SPICE input[3] including subcircuits.

Results

Loop Response. LEX and CEX are set to 10**9h and 10**9f to perform an ac open loop analysis at the closed loop dc operating point. This adjustment opens the loop for ac analysis. Source VEX injects in 1 Vac signal at the duty ratio input of the power stage. The loop gain is the voltage at the node 15, the output of control IC and is shown in Fig. 4 along with Middlebrook's measurements.

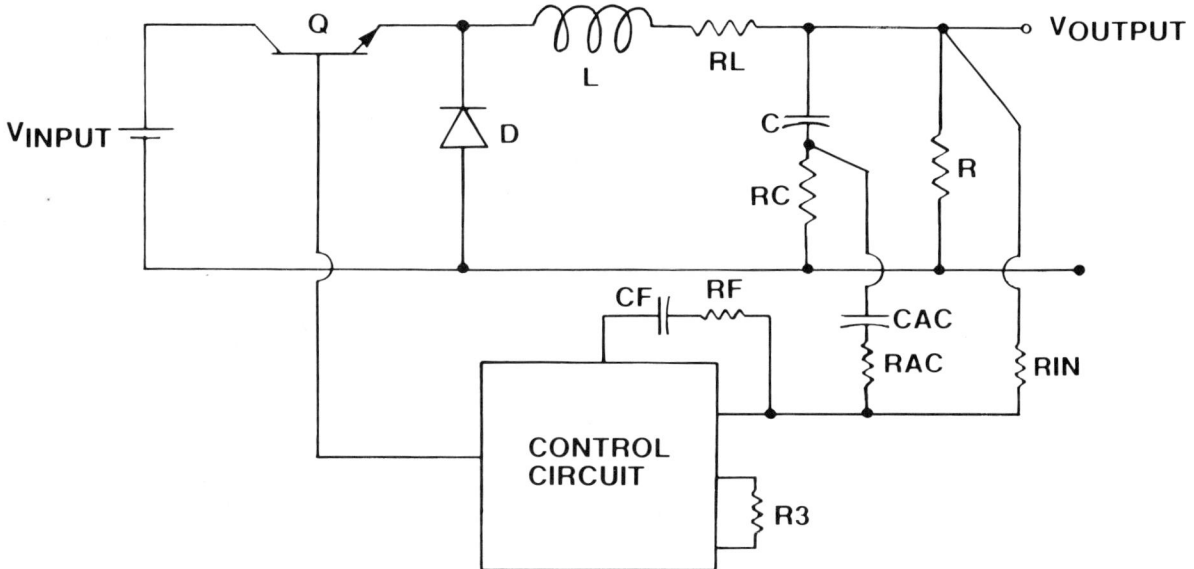

Fig. 1. 10 volt, 1 amp switching regulator circuit.

The cross-over frequency and the phase margin match very well.

Effect of Low Frequency Gain on Dc Regulation. CF and RF control the low frequency gain once the input resistance or the impedance is fixed. Also note that the frequency of the "ZERO (lead)" depends upon CF and RF. Effect on output voltage as a function of CF value is shown in Table 4-1.

Output Impedance. Delete VEX and CEX.

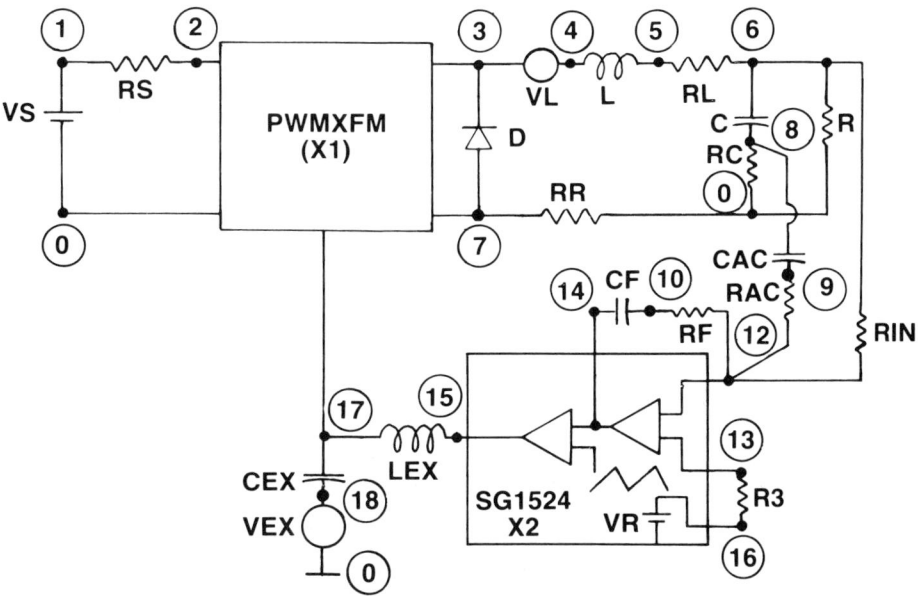

Fig. 2. Equivalent circuit of Fig. 1 ready for SPICE-2 input.

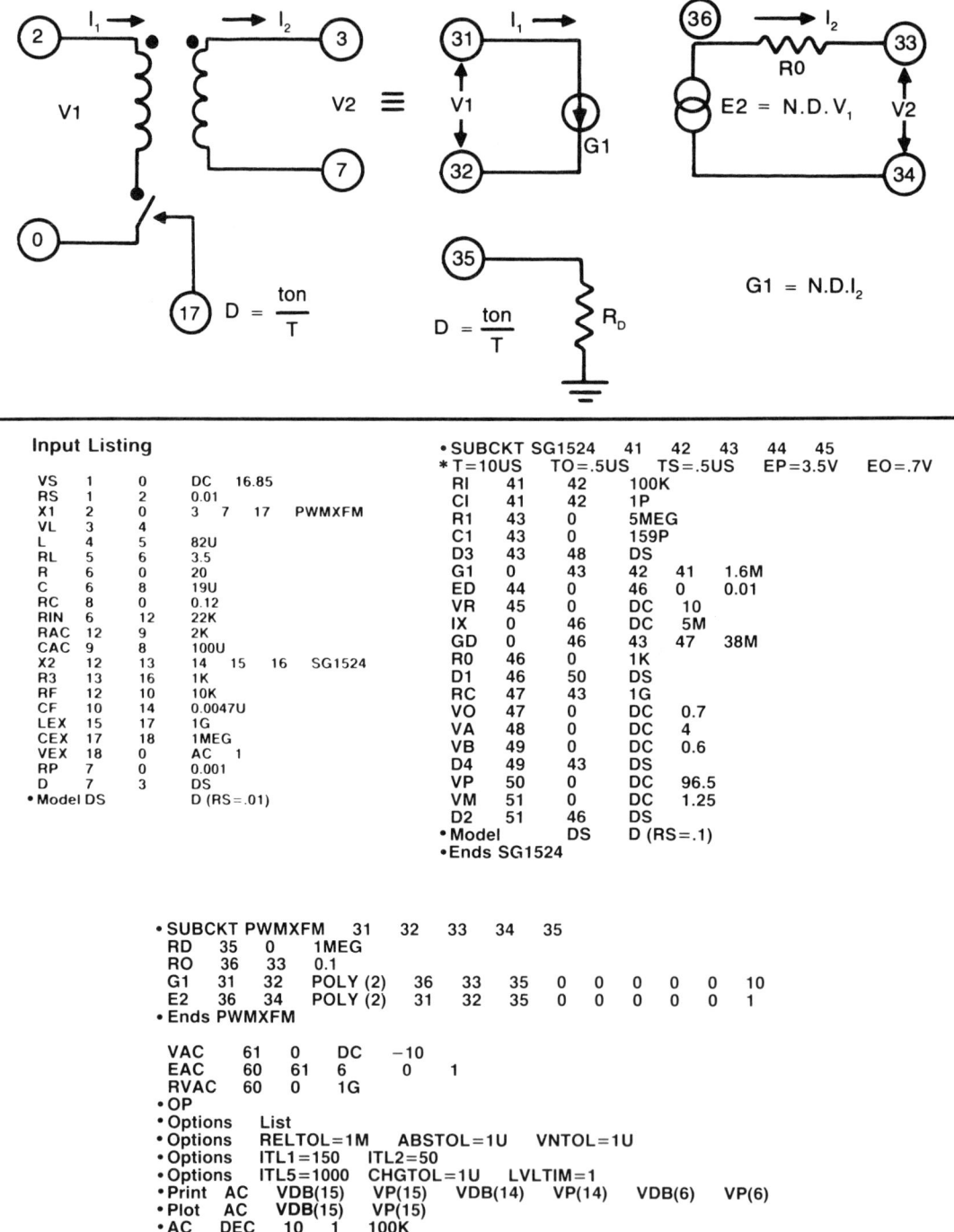

Fig. 3. SPICE input for Fig. 2 including subcircuits.

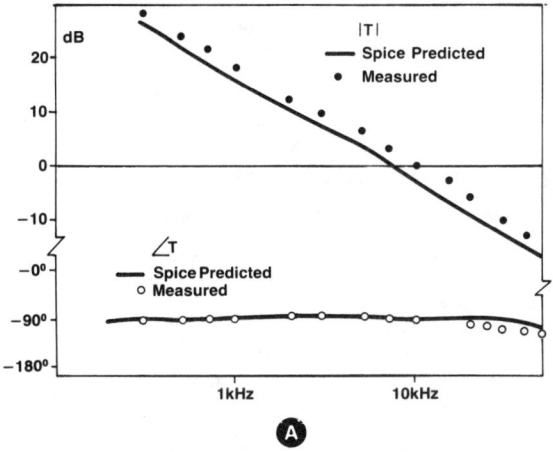

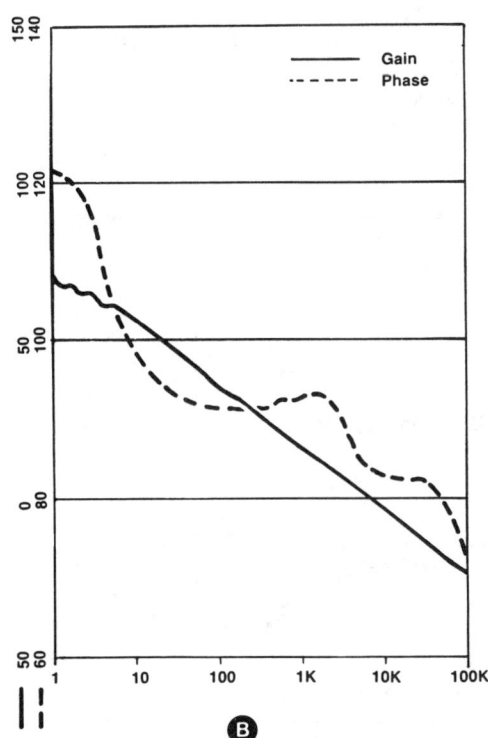

Fig. 4. Open loop gain and phase vs frequency.

Table 4-1. CF Values and Output Voltages.

CF Value (micro F)	Gain at 1 Hz (dB)	Output is regulated at (Volts)
0	−112	13.8823
0.0047	70	9.9996
0.047	36.5	9.9996
47	5.95	9.9996

Add an ac voltage (VOSC) source as shown in Fig. 5A, ac coupled with a large capacitor (CL =) and a zero voltage source VNAM. Figure 5B shows the open loop output admittance. As VOSC = 1V, current through VNAM is the admittance in mhos. For closed loop output admittance, in addition to deleting VEX and CEX, change LEX = 1P. Now the current through VNAM is the closed loop output admittance in mhos. From the admittance, impedance can be computed easily. Figure 5C shows predicted Vs measured closed loop output impedance. Figure 5D shows SPICE output of closed loop admittance as it is.

Input Impedance. An ac voltage source (VOSC) is added in series with the existing input dc voltage (VS). A zero voltage ac source (VNAM) is also added as shown in Fig. 6A to measure the input current. Delete VEX and CEX for open loop

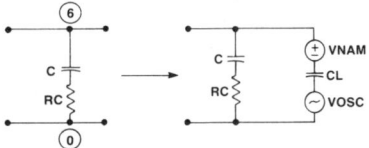

Fig. 5A. Addition of ac voltage source to the output.

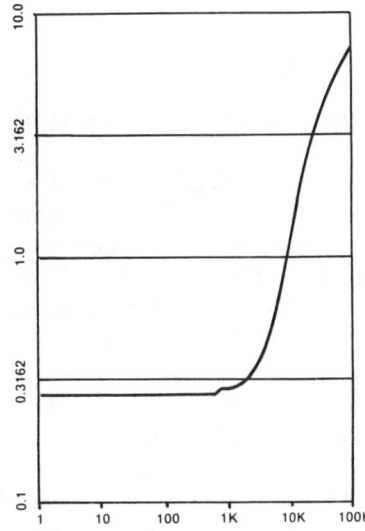

Fig. 5B. Open loop output admittance Vs frequency.

87

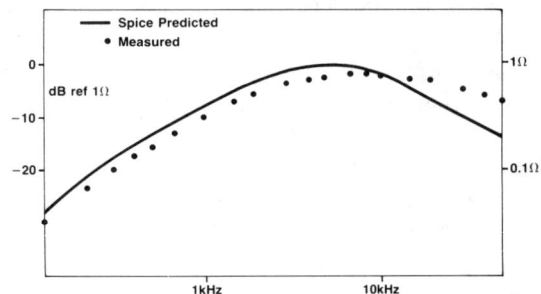

Fig. 5C. Predicted and measured closed loop output impedance.

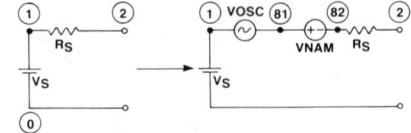

Fig. 6A. Addition of an ac source to the input.

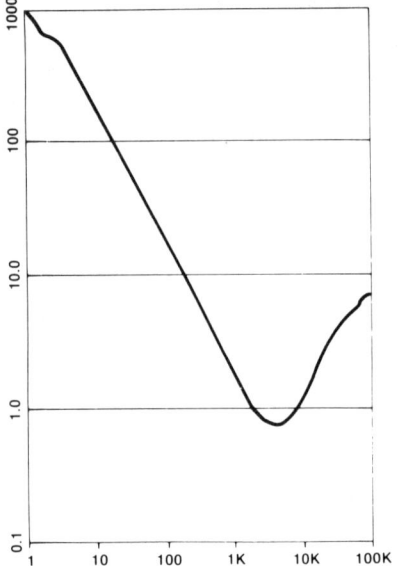

Fig. 5D. Closed loop output admittance vs frequency.

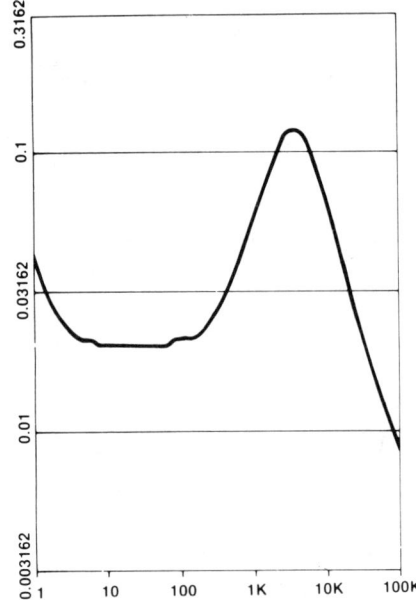

Fig. 6B. Open loop input admittance vs frequency.

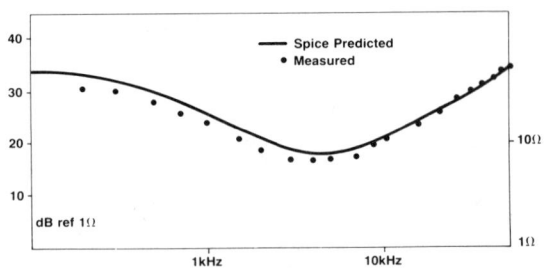

Fig. 6C. Predicted and measured open loop input impedance.

input admittance prediction. As VOSC = 1V, the current through the VNAM is the input admittance in mhos. Figure 6B shows the open loop input admittance (output of SPICE as it is). Now the input impedance can be computed easily from input admittance. Figure 6C shows the open loop input impedance along with measured values. For closed loop input admittance in addition to deleting VEX and CEX, change LEX = 1P. Now Fig. 6D shows the closed loop input admittance.

Line Transmission Characteristic, F. To predict line transmission characteristic, an ac voltage source (VOSC) is added in series with input dc voltage source as shown in Fig. 7A. Delete VEX and CEX for open loop line transmission characteristic, F. As VOSC = 1V, the voltage at node 6 gives the open loop line transmission characteristic and is shown in Fig. 7B. For closed loop line transmission characteristic in addition to

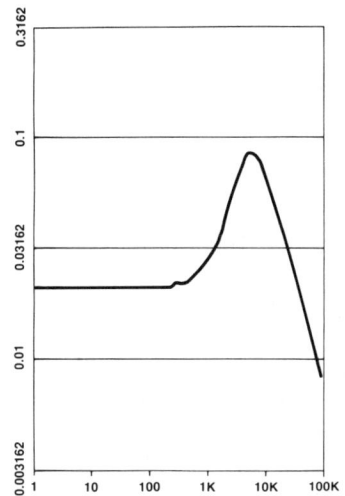

Fig. 6D. Closed loop input admittance vs frequency.

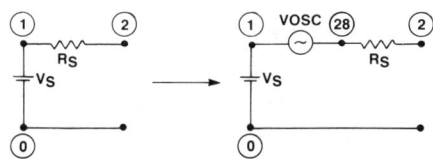

Fig. 7A. Addition of an ac voltage source to the input.

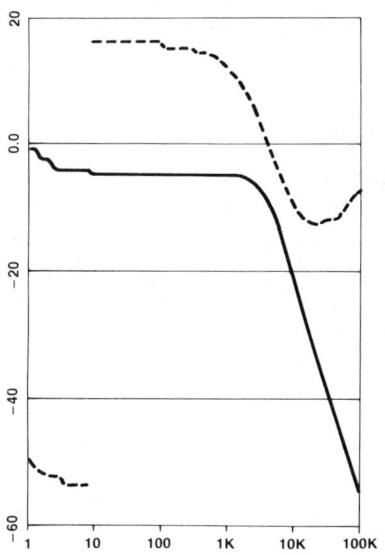

Fig. 7B. Open loop line transmission characteristic vs frequency.

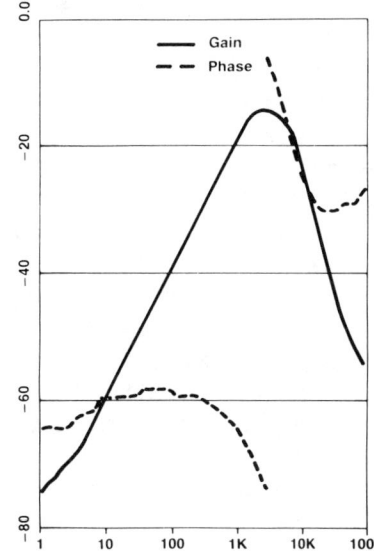

Fig. 7C. Closed loop line transmission characteristic vs frequency.

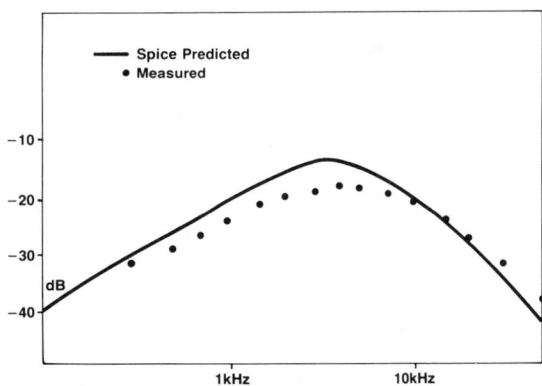

Fig. 7D. Predicted and measured closed loop line transmission characteristic.

deleting VEX and CEX, change LEX = 1P. Now Fig. 7C shows the closed loop line transmission characteristic of the SPICE output as it is, whereas Fig. 7D shows the predicted and measured closed loop line transmission characteristic.

Ouput Load Step Response. Delete VEX and CEX. Change LEX = 1P. Delete "AC" card and insert "TRAN" and "IC" cards.

```
.TRAN 10U 1000U    0U 1U IC
      6    0    PWL(0 0 10U 0.2)
```

Figure 8A shows the output voltage response to a step load of 0.2A. The output voltage dips by 1.26%, overshoots by 0.8% and settles in 370 microsec. Figure 8B shows the control voltage response to a step load. The steadystate duty ratio of 0.595 increases to 0.6316, dips to 0.585 and settles in 370 microsec to a new steadystate value of 0.5974. Voltage drops are proportional to current and hence duty ratio has to increase by a small amount to meet new current demand, while regulating the output voltage at the same value. A vice-versa will happen when the load decreases.

Now the compensation is modified to achieve large bandwidth by using the control scheme shown in Fig. 9A. Figure 9B shows the loop gain and phase vs frequency. It has a bandwidth of 10 kHz with a phase margin of 45 degrees and gain margin of more than 30 dB. Figure 9C shows the step response i.e., undershoot of 1.15%, overshoot of 1.1% and settles in 860 microsec. This indicates that although the bandwidth has increased, the transient

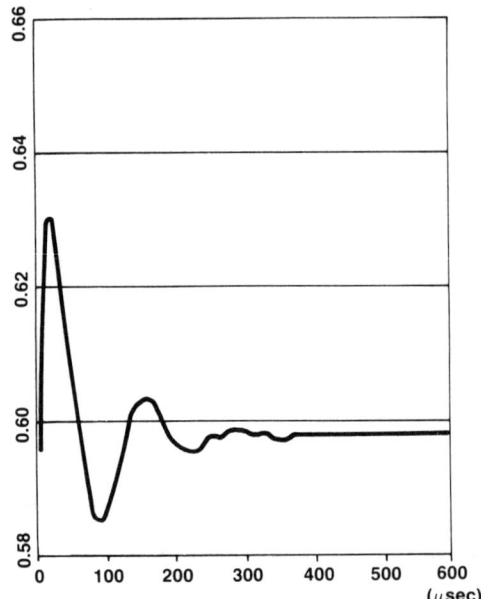

Fig. 8B. Control voltage response to a step load.

response has not been improved. This is because output impedance has a peak. In such a case, the system has to be damped. Thus, the response can be improved by damping the filter (Fig. 10A) and Fig. 10B shows the step load response for damped system. The undershoot is 0.94%, the overshoot is 0.6% and settles in about 290 microsec. Thus the load response has improved significantly.

Conclusions

An efficient and highly useful computer aided design has been presented in this paper which simulates the nonlinear switching dc-dc converter to study its properties of principal interest in the design of modern switch-mode power supplies.

For detailed practical measurement of properties of switch mode dc-dc converter regulators, see

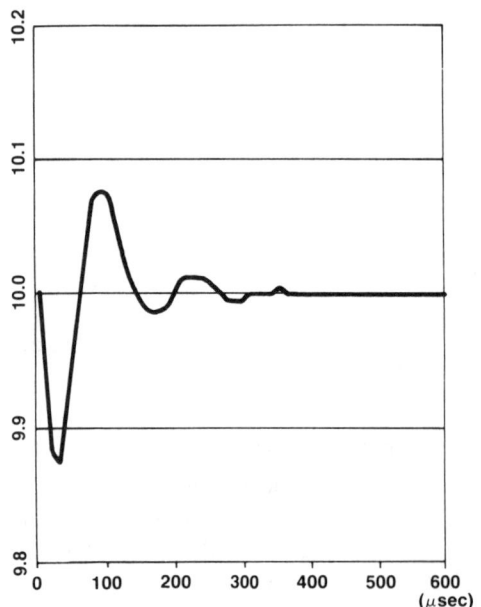

Fig. 8A. Output voltage transient response to a step load of 0.2 A.

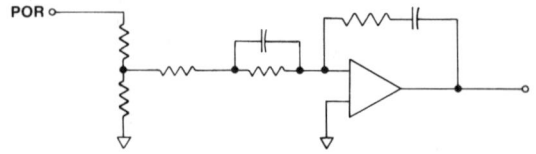

Fig. 9A. Modified compensation control scheme.

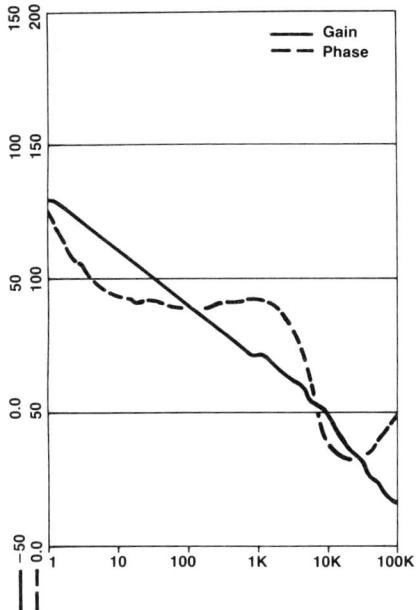

Fig. 9B. Loop gain and phase vs frequency.

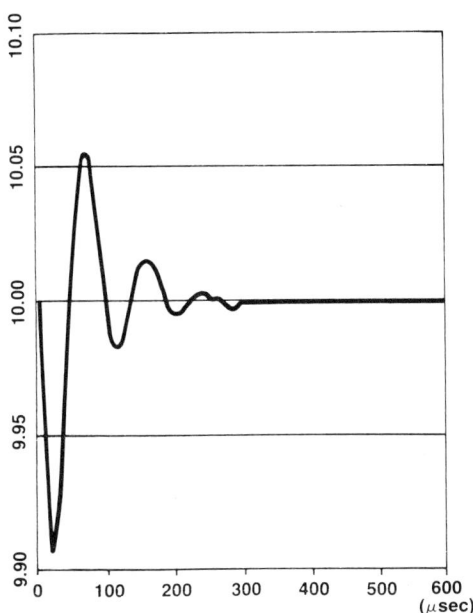

Fig. 10B. Step load response with a damped filter.

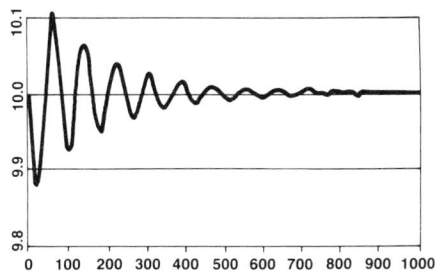

Fig. 9C. Output response to a step load with modified compensation network.

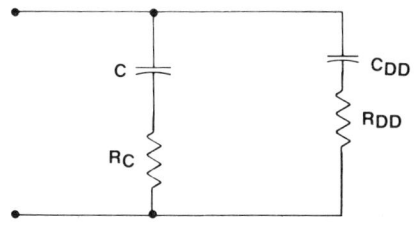

Fig. 10A. Filter damping network.

the reference[4]. Reference[3] dealt with switching regulator analysis whereas the present paper provided a SPICE-2 CAD package to analyze and design the power supplies more efficiently.

References

1. P.R.K Chetty, "Modeling and Design of Switching Regulators," *IEEE Transactions of Aerospace and Electronic Systems*, Vol-AES-18, No. 3, May 1982.
2. R.D. Middlebrook, "Input Filter Considerations in Design and Applications of Switching Regulators", IEEE Industry Applications Society Annual Meeting, 1976 Record, pp. 366-382.
3. Vincent Bello, "Computer Program adds SPICE to Switching Regulator Analysis," *Electronic Design*, March 31, 1981, pp. 89-95.
4. P.R.K. Chetty, "Closed Loops—On Track for Testing Switchers," *Electronic Design*, July 7, 1983, pp. 135-140.

Chapter 5

Practical Design Examples

Design of a 2.8 kW Off-Line Switcher Using PWM Push-Pull Converter   94

Microprocessor-Controlled Digital Shunt Regulator   105

Multiphase Operation of Self-Oscillating Switching Regulator   119

Dc-dc Converter Maintains High Efficiency   123

Linear Power Supplies   127

Improvements to Power Supplies   128

DESIGN OF A 2.8 kW OFF-LINE SWITCHER USING PWM PUSH-PULL CONVERTER *

Design of an off-line 2.8 kW switcher employing pulse width modulated push-pull converter as power stage is presented including step by step design procedure which eliminates trial and error approach and results in small man-hours spent in the development. This regulator presently working in our electronics laboratory has exhibited good stability with a phase margin of 65 degrees while transforming 165 Vdc (rectified from a single phase line) into 280 Vdc isolated output. It has a line regulation of less than 0.5% over ± 20 V input change and a load regulation of 0.5% for a load change from 800 W to 2500 W. This switcher has high efficiency of 90% and high power/weight ratio of 560 W/lb with water cooling.

1.0 Introduction

The off-line switcher takes an important place in power processing electronics and has received considerable attention because of its superior performance, i.e., it eliminates the use of a huge low frequency transformer, exhibits very high efficiency, has high power/weight ratio and offers high reliability. Such an off-line switching regulator employing a pulse width modulated push-pull converter as the power stage has been designed and built to transform unregulated 165 Vdc into a regulated 280 Vdc, delivering a power of about 2.8 kW. Single phase ac line voltage (110 V) is rectified and filtered resulting in 165 Vdc unregulated input voltage to the switcher. The switcher contains the push-pull converter power stage, isolated secondary, rectifiers, two stage LC filters, flux imbalance correction circuit, feedback control circuitry, regulating pulse width modulator, optoisolators, and base drivers.

The following are the important specifications of this off-line switcher:

Input	Single phase ac 110 V ± 15% 60 Hz
Output	Isolated regulated dc 280 V
Power	2.8 kilowatts max.
Regulation	Better than 1% for a line change of ± 20 V
	Better than 1% for a load change from 1.0 kW to 2.5 kW
Ripple	Less than 2% (voltage) over complete operating range
Protection	Over current and short circuit protection
Start Up	Ramp, no large inrush current is expected

The ac input after rectification becomes unregulated dc input to the switcher with 120 Hz ripple components. An off-line switcher as per the above specifications has been designed, built and tested. This switcher has shown expected performance and thus resulted in this paper.

Thus, the main goal of this paper is to present the design of the off-line switcher including the experimental results. After a brief introduction in Section 1, the complete system is described in Section 2. Section 3 presents design of various building blocks, i.e., power stage transformer, rectifier diodes, inductors, turn-on and turn-off snubber, etc. Design of various building blocks will be dealt with step by step procedure which avoids trial and error approach and results in fewer man hours spent in the overall development. Section 4 presents the selection of the various components, i.e., power transistors, diodes for rectification and for snubbers, regulating pulse width modulator, optoisolator, etc. Section 5 deals with the closed loop operation of the push-pull converter to maintain the isolated output at a constant voltage. Experimental results are presented in Section 6, and the final section gives helpful conclusions. It is hoped that

---

*This paper is co-authored by P.R.K. Chetty, Mirza A. Beg, John Dhyanchand and Don Fair.

This work was supported in part by subcontract of N00024-79-C-6022 from Hughes Aircraft Systems and in part by Sundstrand Project Division.

Reprinted with permission from the *Proceedings of the fourth international PCI/Motorcon Conference*, held in March 29-31, 1982, in San Francisco, CA. Copyright © 1982 by Intertec Communications, Inc.

this paper will result in helpful reference for power system designers.

2.0 Description

Figure 1 shows the block schematic of the 2.8 kW off-line switcher using pulse width modulated push-pull converter power stage. It consists of a bridge rectifier to convert single phase 110 V 60 Hz ac into dc (VIN) and a capacitor filter to filter the ripple content. This is fed into a pulse width modulated push-pull converter power stage which transforms unregulated dc at the input square wave ac at the output of the push-pull transformer to a higher voltage level. This ac square wave is full wave rectified and filtered using two stage second order (LC) filters to result in a pure dc. This dc voltage (VO) is scaled down and is compared with a stabilized reference source (VR). The error voltage is amplified, compensated for stable operation of the regulator and fed into a pulse width modulator. The pulse width modulated (PWM) output is used to generate bi-phase PWM signals which drive the alternate power switching transistors, first processing through optoisolators to get isolation and then amplified by the base drive stage. Thus, the isolated output (VO) is regulated at a predetermined level.

The push-pull converter power stage has been selected for the switcher because of the following reasons: A) Output needs to be isolated. Complete input to output isolation is achieved by using optoisolator in addition to push-pull transformer. B) Output voltage is higher than the input voltage. C) The transformer becomes smaller as the positive and negative portions of the BH hysteresis loop has been completely utilized. As the flux swing in the transformer core is in both directions, the transformer design becomes more efficient. D) It can use relatively low power devices. E) This configuration reduces the output ripple by doubling the current ripple frequency to the output filter. Depending upon the availability of switching power transistors and diodes at these power levels, a switching frequency of 50 kHz (referred to the

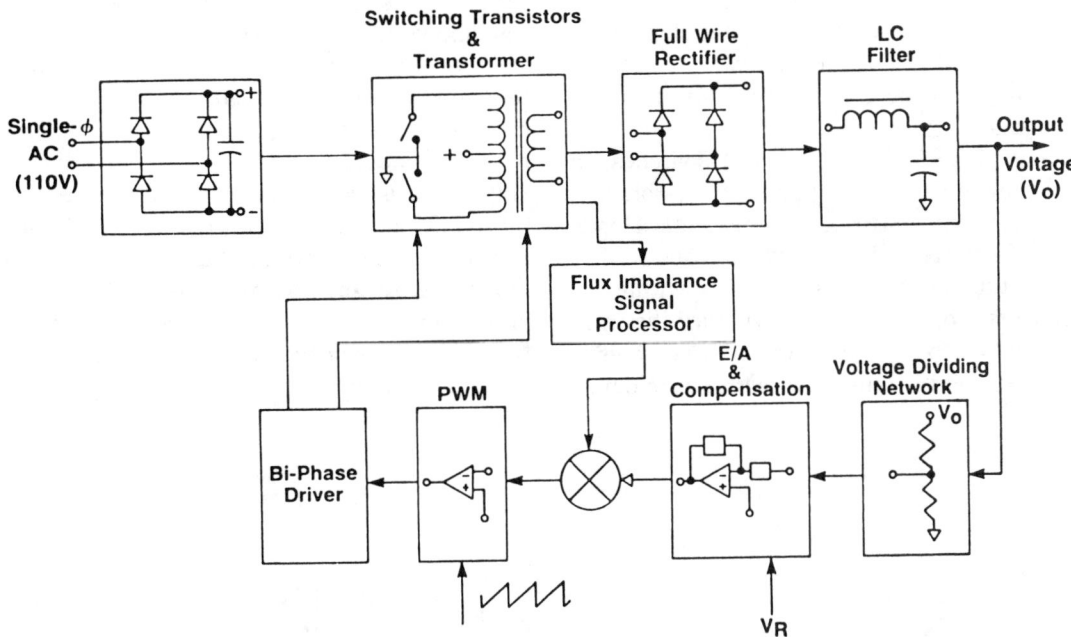

Fig. 1. Block schematic of the 2.8 kW Off-line Switcher. Under specifications the input is mentioned as single phase ac 110 V, 60 Hz. In fact, an alternator is used as input source which had some series inductance always. Hence no physical inductance is shown in rectification and filtering.

secondary of the transformer and 25 kHz referred to the BH hysteresis curve of the transformer) is selected. Further, frequencies below 20 kHz produce audible noise in the form of hum.

A pulse width modulated push-pull converter is a two phase buck converter with isolation. [1] The dc and ac relationships and behavior of the pulse width modulated push-pull converter is the same as that of a buck converter but with the transformer primary to secondary turns ratio. As there will be some voltage drop across the rectifier diodes at these power levels, the secondary voltage of the transformer shall be more to compensate such voltage drops.

Hence
and
$$V_O = V_S - V_D$$
$$V_S = V_P \cdot D \cdot (N_S/N_P)$$

where V_P is the voltage across transformer primary, V_S is the secondary voltage, V_D is accounted for diode and other voltage drops, N_S is the secondary number of turns and N_P is the primary number of turns. N_S/N_P is calculated using the values of V_O, V_P and V_D. Now depending upon the ratio of N_S/N_P and the value of D, the output voltage can be greater or smaller than the input voltage.

Though Fig. 1 shows the ideal switches, in practice they are switching power transistors and are the most highly stressed components. Hence they shall be protected from overstresses so that their operating point always lies safely within the respective safe operating areas whether the transistor is forward biased or reverse biased. This is taken care of by using appropriate turn on and turn off passive snubbers.

Before the transistor is turned on, current through the transistor is ideally zero and voltage across the transistor is the input voltage. When the transistor is turned on, the current through the transistor builds up faster than the voltage decrease across the transistor finally to reach saturated collector emitter voltage. In this transition, large voltage and current are simultaneously present which stress the transistor. To alleviate this problem a turn-on snubber (Fig. 2A) is added which consists of a series inductor in series with the transistor. Now, when the transistor is turned on, the series inductor opposes any current change through itself, thereby delaying the current build-up. This allows the transistor to reach saturated voltage level before its current raises appreciably. Thus simultaneous occurrence of large current and voltage is avoided.

When the transistor is on, the current through the transistor is maximum and the voltage across its collector-emitter is minimum. But when the transistor is turned off, the transistor collector voltage raises fast before its current can reduce to zero resulting in simultaneous occurrence of large current and voltage. This is reduced by using turn-off snubber (Fig. 2B) which is a RCD polarized snubber. Now when the transistor is turned off, collector voltage tries to raise, the snubber diode forward biases, capacitor holds the voltage and delays the voltage raise accepting some of the collector current. Thus simultaneous occurrence of large current and large voltage is reduced. In addition, some of the energy normally dissipated in the transistor is stored in the capacitor. Subsequently, when the transistor turns on again, the capacitor discharges through the resistor dissipating the stored energy.

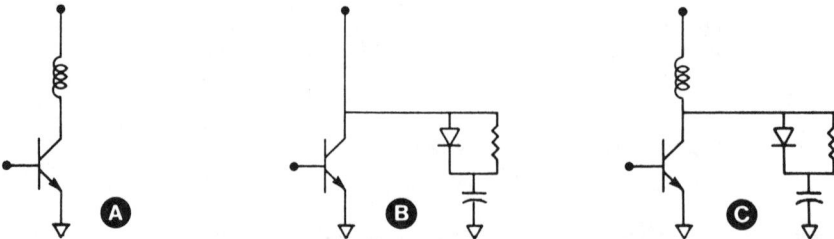

Fig. 2. Transistor switch snubber network.

As the push-pull transformer is dc coupled, most often the core of the transformer walks into saturation at high power and voltage levels. Hence, a flux imbalance correction circuit is employed to protect from this problem. This regulator in addition to above features has over-current and short circuit protection. Also it has a smooth startup circuit such that it does not demand large inrush currents which requires large input filter capacitor.

A steady state duty ratio of 70% is used in the design in view to achieve less harmonic generation in the output. The Fourier analysis shows that: the higher the duty ratio—the lower the harmonics generated, and filtering becomes easy. Though a higher duty ratio might have been selected, to allow enough flexibility for better transient response, 70% is selected. The off-line switcher which uses a PWM push-pull converter power stage is designed to operate in continuous inductor conduction mode over complete output power range and is duty ratio programmed. With the specifications included in the introduction the design of various building blocks is dealt with below:

3.0 Design of Off-Line Switcher

Various building blocks are considered one after another.

3.1 Transformer

At frequencies above 20 kHz, eddy current losses are much greater than hysteresis losses. This can be minimized by the use of very thin laminated cores of 1 or 2 mil thick. A comparison study is carried out to help in the selection [2,3] of the proper material and the results are presented in Table 1. Following are the design specifications for the transformer core selection:

Power Output	=	2800 Watts
Operating Frequency (F)	=	25000 Hz
Operating Flux Density (B)	=	4 kG
Lamination Thickness	=	1 Mil
Curie Temperature	>	300° C
Minimum Core Losses		
Smaller Size and Weight		

Table 1. Characteristics of Magnetic Materials.

Material By Trade Names	Composition	Saturation Flux Density (KG)	Core Recommended Operating Frequency	Losses w/lb. at 5 KG 25 kHz	Smallest Thickness Available (MIL)	Curie Temperature 0° C.
1. Selectron Magnesil Microsil Supersil	3% Si 97% Iron	15 - 18	60 - 100 Hz	300*	1	450
2. Orthonal Deltamax 49 Square Mu	50% Ni 50% Iron	14 - 16	60 - 8000 Hz	160	1	500
3. 4 - 79 permalloy sq. permalloy 80 sq. mu 79	79% Ni 17% Iron 4% Moly.	6.6 - 8.2	1 kHz - 75 kHz	55	1	460
4. Supermalloy	78% Ni 17% Iron 5% Moly.	6.5 - 8.2	1 kHz - 75 kHz	35	1	360
5. Supermendur	49% Co 2% Va 49% Iron	22 - 24	0.75 kHz - 1.5 kHz	#	2	940
6. Ferrite Cores†	Mn Zn Ni Zn	3 - 5 3 - 5	10 kHz - 2 Mkz 200 kHz - 100 MHz	≈ 88 ≈ 88		300 450

*Extrapolated;
Not available for 4 kHz. Seems losses too high at this flux density & frequency.
† Maximum power handling capabilities of presently available ferrite cores are less than 1 kw per core.

To meet these specifications, it is clear from the comparison Table 1 that Permalloy 80 is optimum choice to keep the core losses to a minimum. Using the formula for power handling = 13.W/B.F, MC0017-1D C-Core has been selected. Now the primary turns (NP) are determined using Faraday's law:

$$NP = (VP \cdot 10^8)/(4B \cdot AC \cdot F)$$

where AC = core cross sectional area in sq. cm. NS is calculated using the primary to secondary turns ratio. Current density is determined using the formula [4]:

$$J = KJ \cdot AP^{-0.14} \; AMP/SQ \; CM$$

where KJ is the constant related to core configuration and is equal to 468 for C-core at 50°C. AP is the area product in CM^2 and is equal to 2.52 CM^2 and the wire gauge is given by:

$$AWG = IRMS/J$$

and

$$IRMS = SQ \; RT \; [I^2/D]$$

Core losses are found from the data, core losses per pound provided by the manufacturer. Details of the transformer are summarized below:

Core MC0017-1D

AC = 3.2 CM^2
VP = 168.5 V
WA = 5 CM^2
NP = 14 turns of a rectangular wire (EQ #9 AWG)
NS = 35 turns of #14 AWG
RP = 0.0033 Ohm
RS = 0.0256 Ohm
Window factor is 51%
Copper losses = 4.12 W
Core losses = 14.4 W
B operating = 3.75 KG

3.1.1 Transformer Winding Technique

During turn off, low leakage inductance helps by reducing voltage stress on power transistor and snubber diode. Whereas, during turn on, high leakage inductance helps by reducing current stress on the power transistor. It is not physically possible to provide variable leakage inductance in the transformer. So, it is better to reduce the leakage inductance of the transformer as much as possible. Then, by means of a separate swinging choke connecting in series it is possible to take care of variable inductance property during "on" and "off" to minimize stress on power transistor and snubber diode.

Close coupling between primary and secondary is necessary to reduce the leakage inductance of the transformer. This is achieved by interleaving both windings and evenly spreading over both legs.

For example, following are the measured leakage inductances for two types of windings by a bridge.

A) Leakage inductance when both windings on the same leg without interleaving (primary first and secondary on top) is 3 μH.

B) Leakage inductance when both windings evenly spread over on both legs with interleaving is 0.9 μH.

3.2 Rectification

Full-wave rectification can be achieved using two configurations, i.e., one consisting of a center tap transformer secondary followed by use of two rectifiers appropriately, and second consisting of a non-center tap secondary followed by the use of bridge rectifiers. Though the second configuration uses more semiconductor devices, the first configuration could not be used because the ratings of the diodes shall be two times the secondary voltage. This comes to 840 V and there are no fast recovery diodes available at this voltage level with a current of around 7 amperes. Hence, full-wave rectification using bridge rectifiers has been selected. In this case the voltage rating of the diode is 420 V only.

3.3 Filter Design

The less the current ripple the less the stresses on the power transistors and diodes. To have a lower output ripple voltage and current, a two stage LC filter is chosen. A maximum current of 13.5 A is assumed as the dc current is 10 A corresponding to 2.8 kW. A voltage drop of 5 volts is assumed across the second filter inductor. Now the value of the first inductor is calculated using the formula:

$$L = VL \cdot dT/dI$$

where VL is the voltage across the inductor, dT is the on duration and dI is inductor peak-to-peak current. Assuming a 3% voltage ripple across the first filter capacitor, the value of the capacitor is calculated using:

$$C = dI \cdot TS / 8 \cdot dVC$$

where dI is the inductor peak-to-peak current, TS is the switching period, dVC is the ripple voltage across the capacitor. Following are the values of the inductor and capacitor:

$$L = 300 \ \mu H \ and \ C = 2 \ \mu F$$

Now let us check whether this value of inductor can keep the system operating in continuous inductor conduction mode. This is true if [2 L/R. TOFF] is greater than 1. This constant came to 1.5 at minimum load and 3.2 at full load.

3.3.1 Core Material Selection

Commonly ferrite and molypermalloy powder (MPP) core materials are selected for filter inductors used in pulse-width modulated (PWM) switching regulators. MPP cores operating with a dc bias of 0.3 tesla have only about 80% of original inductance with very rapid falloff at higher densities. Compared to MPP cores, C-cores and cut cores fabricated from grain-oriented silicon steel have approximately four times the useful flux density capability while retaining 80% of the original inductance at 1.2 tesla. Silicon steel cores also provide greater flexibility in the design of high frequency inductors because the air gap can be adjusted at any desired length and because the relative permeability is high even at high dc flux density. Use of these cores will result in smaller size and weight. However, the design of an inductor also frequently involves consideration of the effects of its magnetic field on other devices near where it is placed, as it can be picked up by the nearby circuit in highly dense electronic package. For this type of design problem it is frequently imperative to use a toroidal core. The magnetic flux in a MPP toroid can be contained inside the core more readily than in a C-type core, as the winding covers the core along the whole magnetic path length.

Ferrite E cores and pot cores offer the advantage of low cost and low core losses at high frequencies. However, there are no cores commercially available for kilowatt range power applications. Hence, MPP core has been selected for the present application.

3.3.2 Inductor Design

First, the energy handled by the inductor is calculated using the formula LI^2 where L is the inductance required with dc bias and I is the maximum dc current through the inductor. A core material of molypermalloy powder is chosen for this application as mentioned above because of its distributed air gap which supports large dc bias. Now using the dc bias core selector chart [5] and as highest permeability of 60 μ is selected. The number of turns to get the required inductance is calculated using the formula:

$$N = 10^4 \sqrt{\frac{L}{L_{1000} \times \% \ permeability}}$$

where L = desired inductance in μH at no load, L1000 = nominal inductance in μH per 1000 turns, and % permeability is obtained from permeability vs dc bias curves [5]. Dc bias or magnetizing force is calculated using the well known Ampere's law.

$$H = \frac{0.4 \ \pi \ NI}{1}$$

where
- H = magnetizing force in oe
- N = number of turns
- I = peak magnetizing current in amperes
- l = mean magnetic path in CM

From the above two equations it can be seen that % permeability is a function of H and H is a function of N, which is to be calculated. Hence, these two interdependent equations have been solved to calculate N. Assuming a window factor (WF) of 40%, knowing the core window (WC) in circular mils, value of N, wire area (AW) is calculated in circular mils using the formula:

$$WF = (AW \cdot N)/WC$$

Core losses are determined using core loss chart which permit calculation of core loss in ohm per millihenry and in watt per pound. Thus losses in watt per pound is given by:

$$Pc = 4 B^2 R \times 10^{-6}$$

$$B = \frac{Erms \times 10^8}{4 A_c N f}$$

$$Erms = \sqrt{D(Vl^+)^2 + (1-D)(Vl^-)^2}$$

where
- Vl^+ = Voltage across inductor during switch ON time
- Vl^- = Voltage across inductor during switch OFF time

Following similar procedure the second stage filter is also designed and details of filter inductors are summarized below:

Inductor 1
55083, 60 μ MPP Core
M type Temperature Stabilized
WA = 842700 CIR MILS
AC = 1.072 Sq. CM.
LM = 9.84 CM.
Core Wt = 0.206 lbs.
N = 85 of #14 AWG
Core Losses = 2.3 W
Copper Losses = 4.3 W
Permeability 49% at Full Load
BAC = 910 Gauss
L = 585 μH with no dc bias
Inductor Weight = .39 lb.

Inductor 2
55083, 60 μ MPP Core
M type, Temperature Stabilized
WA = 842700 CIR MILS
AC = 1.072 Sq. CM.
LM = 9.84 CM.
Core Wt = 0.206 lbs.
N = 60 of #14 AWG
Core Losses = Negligible
Copper Losses = 2.7 W
Permeability 55% at Full Load
BAC = 40 Gauss
L = 292 μH with no dc bias
Inductor Weight = .34 lb.

3.4 Snubber Diode Design

The snubber diode in the push-pull converter provides a low impedance path to the snubber capacitor and thus performs an essential function of allowing the turn-off current to flow through the capacitor when the transistor (Q1) turns off. The voltage at the anode end (D1) gradually increases and reaches input voltage. When the opposite transistor (Q2) turns on, the voltage at the anode end (D1) becomes twice the input voltage because of auto transformer action. Further, in the practical transformer, leakage inductance is not ideally zero resulting in a voltage peak ranging from 2.1 to 2.7 times the input voltage. Thus the diode shall be rated for this high voltage and to a peak current rating equal to the current through the transistor just before its turn off.

3.5 Snubber Design

It should be noted that the snubber designed to reduce transistor stresses during turn on tends to increase the stress during turn off, i.e., energy stored in the shunt capacitor increases the turn-on current and energy stored in the series inductor increases the turn-off voltage.

A "normal" size turn-on inductor is defined [6] as that which allows the transistor current to reach its final value at the same time as the voltage reaches zero. The "normal" size of turn-off capacitor is also similarly defined.

As per the definition, the normal sizes are:

Inductance $LN = VIN \cdot T / (2 \cdot IP)$
Capacitance $CN = IP \cdot T / (2 \cdot VIN)$

where IP is the current just before transistor turns-off.

$$T = 1.0 \ \mu S \ (assumed)$$

It is found that the overall losses will be minimum when losses in switching transistor equals the losses in snubbers [6] and thus the optimum values of L and C are given by:

$$L = 4/9 \ LN \ and \ C = 4/9 \ CN$$

Calculated values of $L = 1.1 \ \mu H$ $C = 0.045 \ \mu F$

Details of the snubber L,C as used are summarized below:

Series Inductor
    Core   55120-M4
    LM   4.1 CM
    AC   0.192 Sq. CM
    Core Weight   0.24 Oz
    H   118 oe
    L   1.3 μH at Full Load
    Number of Turns 11 of #14 AWG
Shunt Capacitors 0.05 μH

3.5.1 Transistor and Snubber Losses

Transistor energy loss during turn on
$E(TR-ON) = VIN \cdot IP \cdot TS/2) (1 - (4/3)$
    $SQRT \ (L/LN) + L/(2 \ LN) \ )$

Transistor energy loss during turn off
$E(TR-OFF) = VIN \cdot IP \cdot TS/2) (1 - (4/3)$
    $SQRT \ (C/CN) + C/(2 \ CN) \ )$

Energy stored in turn on snubber
$E(L) = L \cdot IP^2/2 = VIN \cdot IP \cdot TS/4 \ (L/LN)$

Energy stored in turn off snubber
$E(C) = C \cdot VIN^2/2 = VIN \cdot IP \cdot TS/4 \ (C/CN)$

Power loss in transistor during turn on
    $= E(TR-ON) / TS$
    $= 21.7 \ W$

Power loss in transistor during turn off
    $= E(TR-OFF) / TS$
    $= 22.85 \ W$

Power loss in transistor during on period = 12 W
Total Transistor Loss = 56.55 W

Turn on snubber loss (both sides)
    $= E(L).2 / TS$
    $= 40 \ W$

Turn off snubber loss (both sides)
    $= E(C).2 / TS$
    $= 37.2 \ W$

Total snubber loss = 77.15 W

3.6 Base Drive

The circuit should be simple and reliable. The on and off states shall be ensured with no false triggering due to noise or other effects. A positive current pulse with proper rise time is required for proper switch turn-on and relatively large negative base drive is required for fast turn off. The design and operation is self-explanatory from the circuit diagram shown in Fig. 3. To suppress the line transients, the base drive circuit uses bypass capacitors at appropriate places.

4.0 Component Selection

4.1 Power Switching Transistor

As described above, the power transistor

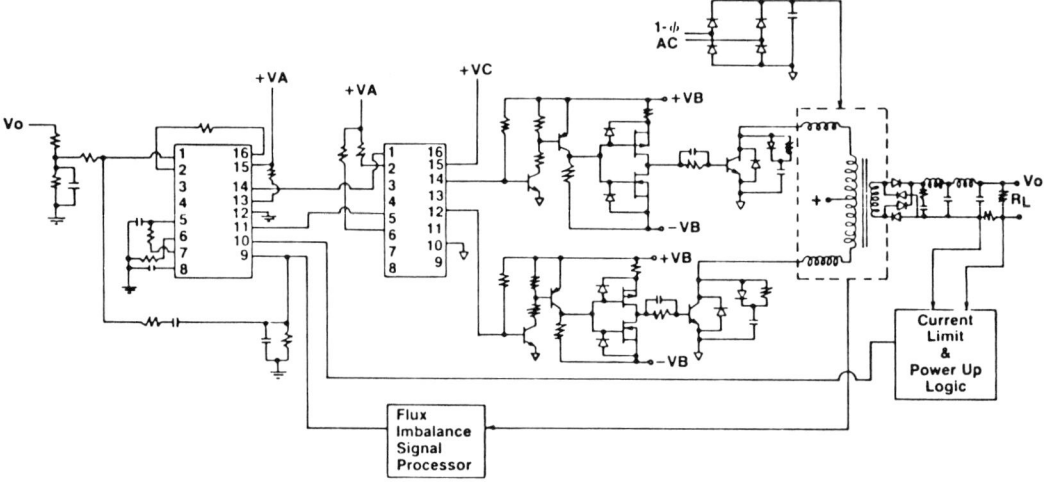

Fig. 3. Detailed circuit diagram of the off-line switcher.

switch is the component which is stressed most. These stresses are of three kinds:

a) When the transistor is off, lower leakage current and input voltage.

b) When the transistor is on, large on current and collector to emitter saturated voltage.

c) When the transistor is turning on and when turning off. This is the transition state, during which large voltage and large current are simultaneously present, stress the transistor to a maximum.

From the design data and survey made to locate a suitable transistor as a power switch, having selected the push-pull configuration for power stage, resulted in the selection of PT4503 PowerTech transistor, which meets all the design requirements.

4.2 Diodes

Motorola MR1376 diode has been selected to meet the requirements of rectifier diodes as well as snubber diodes. Following are the characteristics of MR1376 diodes:

Maximum Current = 12 A

Maximum Reverse Voltage = 600 V

Recovery Time = 300 nsec

4.3 Capacitors

Polypropylene low ESR capacitors with large ripple current capability have been selected for power filter stage. Stabilized mylar capacitors have been used in the control circuitry.

4.4 Control IC

There are many manufacturers who produce integrated circuits for use in switching power supplies. From the past experience of using various control ICs and from the study carried out resulted in concentrating the comparison of such regulating pulse width modulator ICs to only the following four ICs from third and latest generation. They are 1524, 1525/1527, 1526 and 1525A/1527A. Each of these ICs contains all the control circuitry for a regulating power supply. This includes the stabilized voltage reference source, error amplifier, oscillator, pulse width modulator, pulse steering flip-flop, dual alternating switches and shut down provision. Use of blanking pulse to both outputs ensures that there is no possibility of having both outputs on simultaneously during transitions. In addition to these common features, the comparison in Table 2 shows some of the different features each of these four ICs have. From the table we selected 1527A for our application as most suitable IC.

4.5 Optoisolator

6N134 dual channel hermetically sealed optically coupled logic gate has been selected for obtaining complete input to output isolation. This optoisolator is more suitable for the present application compared to other optoisolators. It has 25 mA drive capability (each channel) with a current transfer ratio of about 400% and is of inverting type with a light emitting diode and a unique high gain integrated photon detector. The output of the detector is an open collector Schottky clamped transistor.

This unique dual coupler design provides maximum dc and ac isolation between each input and output.

5.0 Compensation

For a system to be stable, the loop gain shall fall below unity by the time total phase shift of the system reaches 360 degrees. The power stage filter has a corner frequency of 5.2 kHz and gain decreases with −40 dB slope starting from that frequency. Conventional dominant pole frequency compensation is employed[1] to stabilize the switcher. Dominant pole frequency and the overall loop gain are selected such that the loop gain crosses the zero dB line at about 2 kHz with −20 dB slope. A pole-zero pair is placed such that pole starts at zero frequency and zero frequency lies below the dominant pole frequency. This increases

Table 2. Comparison of Regulating Pulse-Width Modulators.

Parameter	1524	1526*	1525/1527	1525A/1527A
Latches PWM to prevent multiple pulses.	No	Yes	No	Yes
Independent dead time control	No - deadtime is a function of timing capacitor	Yes	Yes	Yes
Common mode range of error amplifier	3.4 V maximum	Up to reference Voltage (5.1 V)	Up to reference Voltage (5.1 V)	Up to reference Voltage (5.1 V)
Internal clamp diode & current source for soft start	Not Available	Available	Available	Available
Separate sync terminal	Not Available, but can be synch.	Available	Available	Available
Totempole output	No	Yes	Yes	Yes
Operating frequency	> 100 kHz	1 Hz to 400 kHz	100 Hz to 400 kHz	100 Hz to 500 kHz
Presence of internal current limiting amplifier	Yes. Sensing is possible only in ground line	Digital current limiting with wide current limit common mode range	No	No
Input under voltage lockout	No	Yes	No	No

*This IC has provision for symmetry correction inputs. PWM comparator has hysteresis. TTL/CMOS compatible logic inputs.

the low frequency gain guaranteeing good load regulation and line regulation.

6.0 Experimental Results

Figure 3 shows the detailed circuit diagram of the off-line switcher employing pulse width modulated push-pull power stage. This circuit has been built in the power electronics laboratory and tests were carried out to evaluate its performance, against its design specifications. Figure 4 shows the PWM control signal which is fed into the optoisolator and amplified using base drive stage. Figure 5 gives the base voltage and current waveforms. Figure 6 gives collectors voltage, collector current, and base current waveforms. The base drive voltage has a rise time of 400 nsec. and fall time of 500

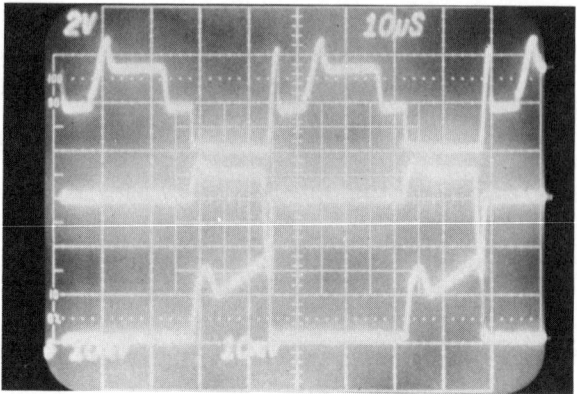

Fig. 6. Collector voltage, collector current and base current waveforms.

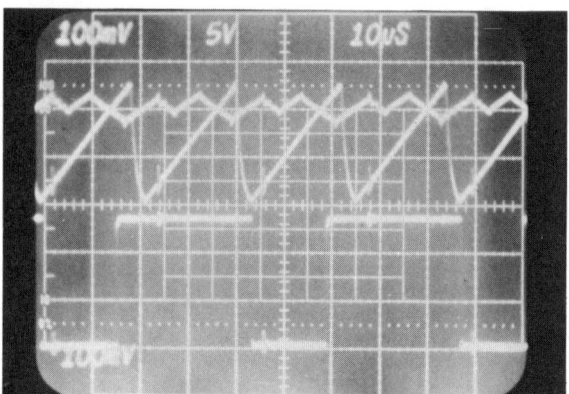

Fig. 4. Top: Compensated amplified error signal; middle: sawtooth ramp; bottom: pulse-width modulated control signal.

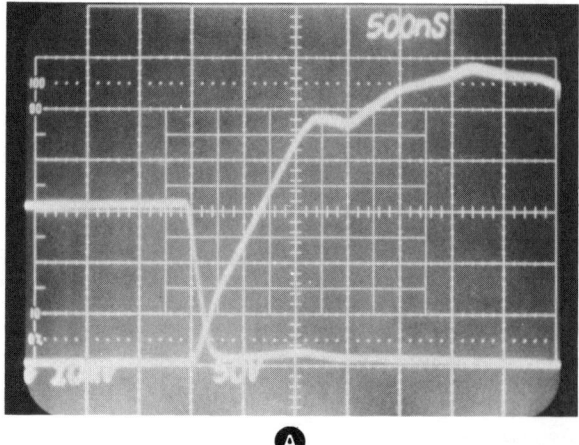

Ⓐ

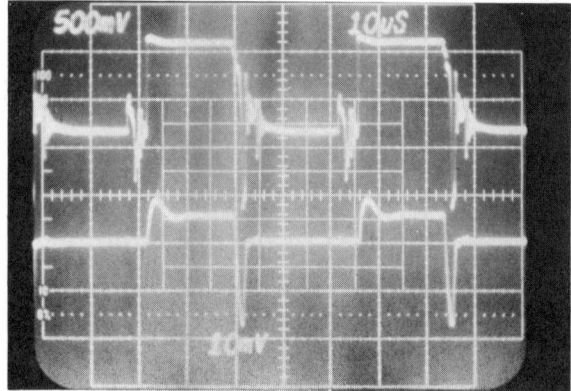

Fig. 5. Base voltage and current waveforms.

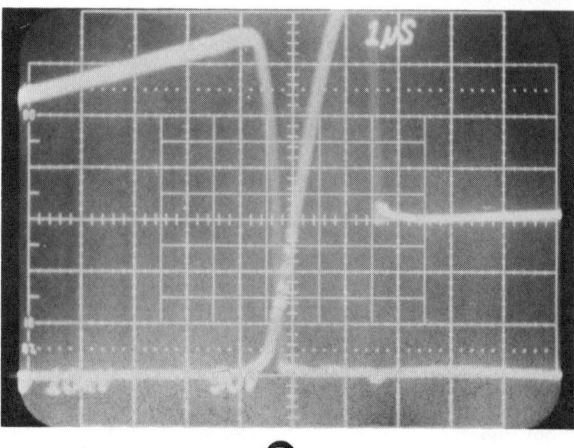

Ⓑ

Fig. 7. Transistor voltage and current, (A) during turn-on and (B) during turn-off.

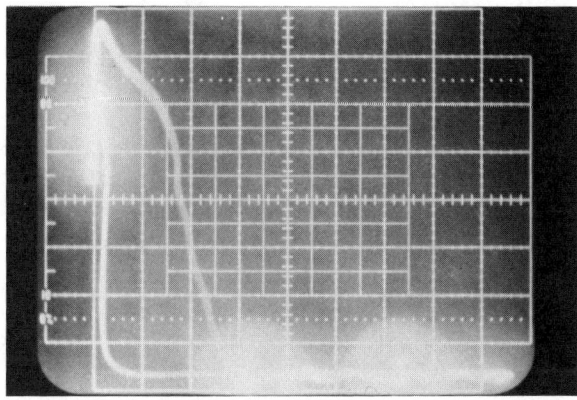

Fig. 8. Safe operating area of the transistor with snubbers.

nsec. The collector voltage and current are affected by the snubbers which protect the transistors. Figure 7 shows the transistor voltage and current during turn-on and turn-off. From this figure it can be seen how effectively the turn-on and turn-off snubbers are functioning. To throw more light into transistor protection their safe operating area is shown in Fig. 8 when the switcher is working. This lies well within the FBSOA and RBSOA provided by the manufacturer.

Following are some of the details of the test results:

Output Voltage	=	280.6 V
Output Power	=	800 W to 2.8 kW
Line Regulation	=	0.5% over ± 20 V input change
Load Regulation	=	0.5% for a load change from 800 W to 2.8 kW

From the above performance details it is clear that the switcher has performed well meeting design expectations.

Conclusion

A 2.8 kW off-line switcher using pulse width modulated push-pull converter is presented in the previous sections in detail. This regulator has an efficiency of 90% at a power level of 2.8 kW and has power to weight ratio of 560 W/lb with water cooling.

The authors of this paper hope that the detailed design procedure presented in this paper will be highly helpful for the power system designers.

References

1. P.R.K. Chetty, "Modelling and Design of Switching Regulator," *IEEE Transactions on AES*, Vol. AES, No. 3, May, 1982.
2. Inverter Transformer Core Design and Material Selection, Magnetics Inc.
3. A Critical Comparison of Ferrites with Other Magnetic Materials, Magnetic Inc.
4. Col. Wm. T. McLyman, *"Transformer and Inductor Design Handbook,"* Marcel Dekker, Inc., New York and Basel.
5. Catalog MPP-303S, Molypermalloy Powder Cores, Magnetics Inc.
6. William McMurry, "Selection of Snubbers and Clamps to Optimize The Design of Transistor Switching Converters," *IEEE Transactions on Industry Applications*, Vol. IA-16, No. 4, July/August 1980.

MICROPROCESSOR-CONTROLLED DIGITAL SHUNT REGULATOR

A new approach to the design of power systems is presented in which a microprocessor is used as a controller for a digital shunt regulator (DSR). This approach meets the demands of future space and ground missions, i.e., high efficiency, high reliability, low weight, low volume, increased flexibility, and less development time. This approach responds to future demands by permitting real-time modification of system parameters for system optimization. This feature is especially important in the event of an anomaly. As the microprocessor need not be dedicated to the DSR, it can simultaneously be used for battery management and for charge regulator/discharge regulator control. This approach also reduces the component count, simplifies assembly and testing of the unit, results in significant time saving, and increases the reliability.

---

*This paper is co-authored by P.R.K. Chetty, W.M. Polivka and R.D. Middlebrook. California Institute of Technology.
© 1980 IEEE. Reprinted with permission from *IEEE TRANSACTIONS ON AEROSPACE AND ELEC-*

1.0 Introduction

In recent years use of microprocessors for power processing systems has resulted in improved performance. To meet the predicted requirements of spacecraft, it is necessary that new and improved methods of electrical power conditioning and control be developed. The power system should be modular and should not take any development time. Other constraints imposed are high reliability, minimum weight and volume, low cost, and flexibility. To meet these demands an attempt has been made to use a microprocessor to control the digital shunt regulator (DSR).

Since the advent of the space age, photovoltaic cell arrays have been used as the main energy source for spacecraft power generation. In addition, solar energy, which is abundant in space, may play an important role in meeting the world energy requirements by means of microwave transmission from space. Because the electrical output of photovoltaic cell arrays is unregulated, however, the power must be processed and regulated before it can be used by other equipment. Thus these solar power systems require bus regulators, of which the DSR is superior compared to other types of shunt regulators [1]. The main advantage of the DSR over other types of shunt regulators is that this can be employed for high-power systems as its weight, size and volume do not increase in proportion to the power requirements as others do. Although the circuitry is somewhat more complex than a simple analog shunt or sequential shunt, it does offer high reliability and is of low cost.

The power systems described above, whether used in space or on the earth must use batteries to meet the peak and eclipse/shadow requirements of the load. Hence, the storage batteries have to be charged during sunlit period/day time and have to be discharged during eclipse/nighttime or when there is a requirement for peak power. In addition, these batteries have to be protected from overcharge and undercharge. Attempts have already been made to use microprocessors for battery management [2,3]. Microprocessors can also be used to control charge and discharge regulators (CR/DR).

Thus the approach of using a microprocessor for the control of the DSR is a very useful one, especially since the same microprocessor may be used for controlling DSR, CR/DR, and battery management. This approach is expected to result in a single integrated system/unit for all the functions mentioned above with a bonus in system flexibility, high reliability, minimum weight and volume, and standardization. The processor can also be used to continuously monitor the status of its own system and the health of the overall power system as well.

The purpose of this paper is to present the hardware and software details of a microprocessor controlled digital shunt regulator.

Section 2 contains a brief description of the digital shunt regulator. Then follows the detailed description of solar array section simulators in Section 3. The operation and design of the dissipative analog shunt is given in Section 4. Section 5 contains the description and design of shunt current comparators. The need for a special timing function and its design is discussed in Section 6. Section 7 contains the description of the microprocessor controller, hardware implementation, software programs, and interfacing. The complete system is described in Section 8. The experimental results of the model system constructed as described above is presented in Section 9. Some possible extensions are suggested in Section 10.

2.0 Digital Shunt Regulator

Figure 1 shows the block diagram of a power system using a digital shunt regulator. As mentioned above, the energy source is a photovoltaic cell array. The digital shunt regulator regulates the output of the energy source to the needs of the load. The solar cell array is divided into N sections, one section of which is permanently connected to the bus and all other sections are connected through

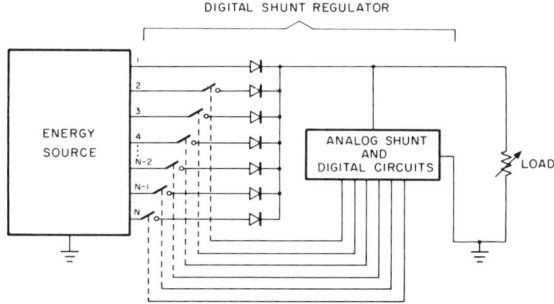

Fig. 1. Block schematic of a power system using digital shunt regulator.

switches. The DSR contains a small dissipative analog shunt which is designed to regulate one section of the array. The current through this dissipative shunt is monitored. Whenever this current exceeds approximately the current of a single section, I_{max}, the digital processing part of the DSR switches off one section. Whenever this current reduces to a minimum, I_{min}, then the digital part of the DSR switches on one section. Thus coarse regulation is achieved by the digital part of the DSR and fine regulation is achieved by the dissipative shunt. This DSR will be described in detail in the following sections.

Because the solar cell array is not available to test the DSR, solar cell array section simulators have been constructed and used instead. First various building blocks of the microprocessor controlled DSR will be explained and then the overall description of the complete system will be given.

The demonstration system has been designed to the following criteria: Regulated bus voltage 28 V, Maximum power to be handled 30 W, Number of solar cell array sections 4.

3.0 Solar Cell Array Simulator

Figure 2(A) shows the I-V characteristic of a typical solar cell array section. For the model system being used here, the simulator need not exhibit an I-V characteristic identical to that of a real solar array—it need only be similar. The nearest simulation with the least complexity is achieved by using a current source (as a solar cell is also a current source) whose I-V characteristic is shown in Fig. 2(B). This I-V characteristic is enough to simulate the solar cell array for testing of the DSR.

Each solar cell array section simulator has been designed to give about 300 mA at 28 V. Figure 3 shows the circuit diagram of one simulator. Four units of this type have been constructed. The I-V characteristics of all four simulators are given in Fig. 4. The portion of the circuit within the dotted line in Fig. 3 is used to switch the simulator on and off. A light-emitting diode (LED) has been included in each circuit as shown to indicate visually whether the section is switched on or off.

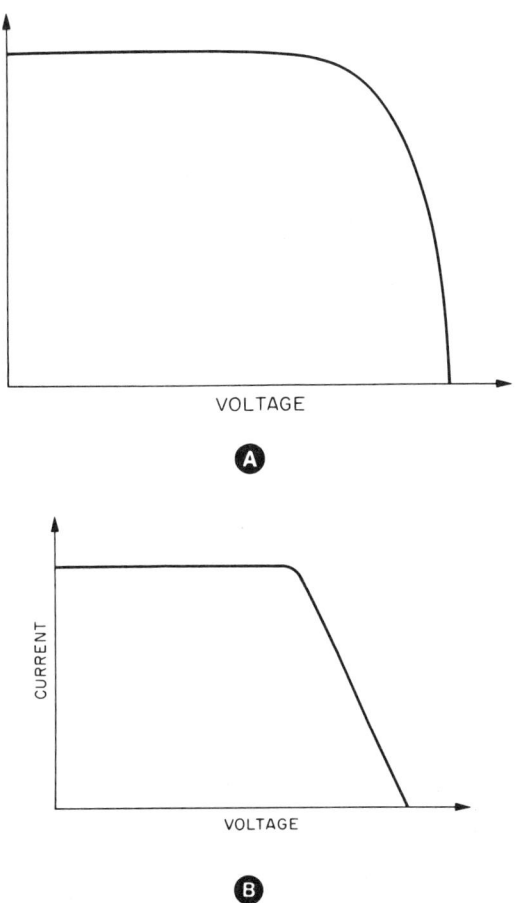

Fig. 2. I-V characteristics of (A) typical solar cell array, (B) solar array simulator.

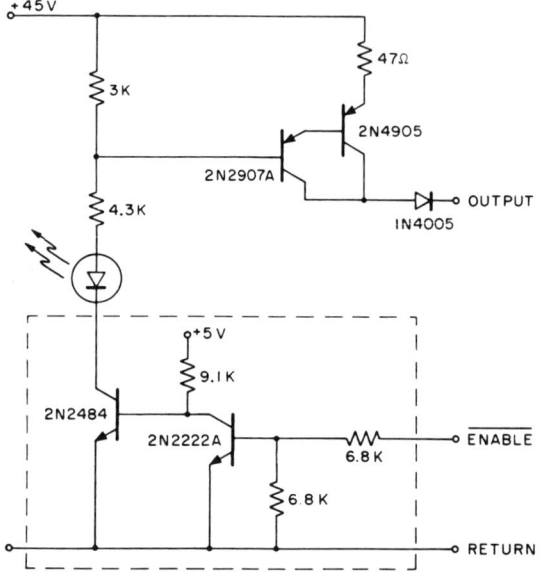

Fig. 3. One section of a solar array simulator.

4.0 Dissipative Analog Shunt Regulator

As mentioned above, the dissipative analog shunt is used to achieve fine regulation of the bus voltage. This type of shunt has been chosen over the alternative pulsewidth modulated shunt because the analog circuit has a much higher bandwidth and therefore is more desirable for achieving finer regulation. Figure 5 shows the circuit diagram of the dissipative analog shunt. Divided-down bus voltage is compared to the reference voltage and the amplified error voltage is used to control the current through the shunt. A resistor of suitable value (depending on the maximum design current of the shunt) is used in the shunt current path to monitor the current that is flowing through the shunt. This signal is used for further digital processing.

5.0 I_{max} and I_{min} Comparators

The shunt current, measured automatically using a resistor in the shunt path (Fig. 5), is compared against two reference voltages to determine whether it is above I_{max} or below I_{min}. Fig. 6(A) shows the circuit diagram of the I_{max} and I_{min} comparators. Figure 6(B) shows the waveforms of the shunt current and the outputs of both the comparators.

The reference voltages for the I_{max} and I_{min} current comparators are derived from the bus voltage by using a zener diode and resistor dividers. It would also be possible to generate the references from digital-to-analog converters. In this way the reference voltages could be changed when necessary. Occasionally this must be done to compensate for degradations due to ageing. In a space system these changes can be telecommanded from ground. In ground-based systems, such maintenance commands can be given from remote locations.

An alternative approach to the one presented above is to use an analog-to-digital converter to put

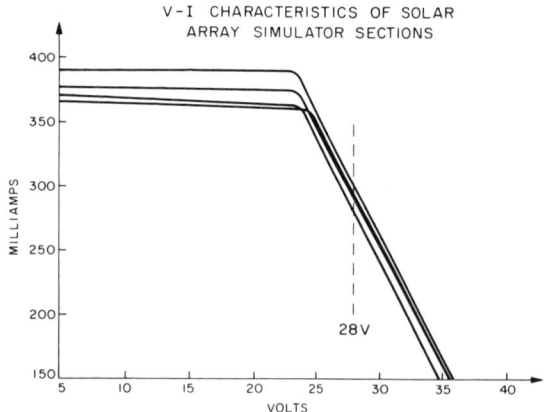

Fig. 4. I-V characteristics of all four solar array simulator sections.

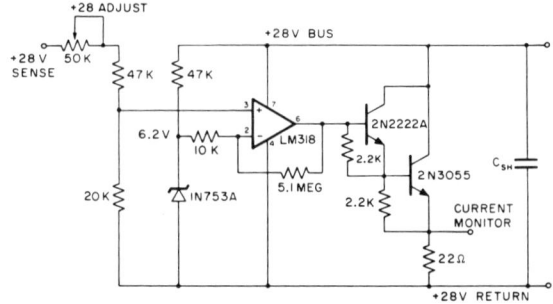

Fig. 5. Analog shunt regulator.

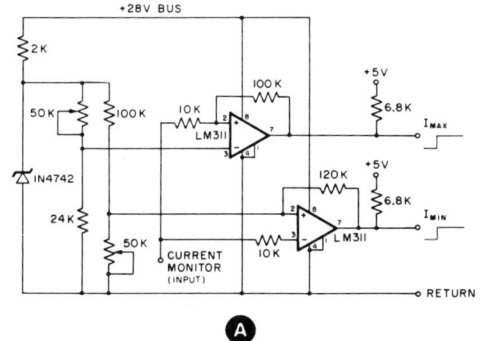

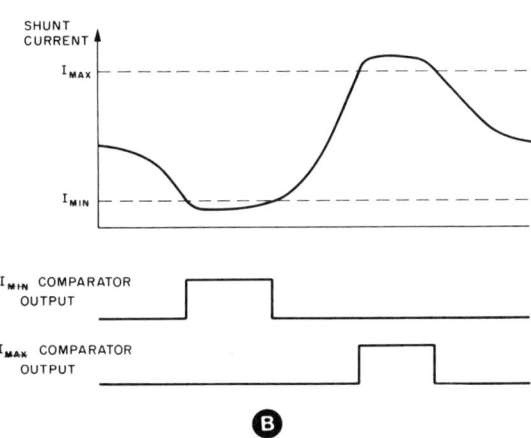

Fig. 6. Current comparators—Relationships between Imin, Imax, and shunt current.

the shunt current signal into digital form. This digital information can then be compared with a digital reference within the microprocessor. This is illustrated in Fig. 7. In this case the digital-to-analog converter for the reference voltages mentioned above is not required.

6.0 Switch Timing

The two outputs from the I_{max} and I_{min} comparators are used by the microprocessor to switch the solar array sections. For stability reasons this is done according to a timing signal from an external clock. This will be discussed in detail in Section 7A.

7.0 Microprocessor Controller

Figure 8 is a general block diagram which

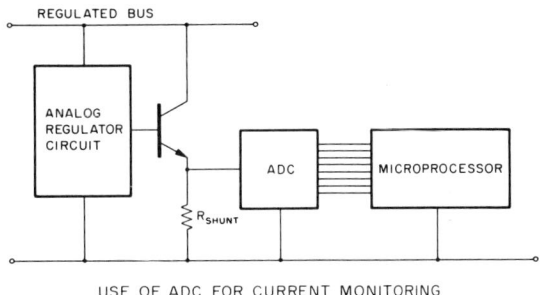

Fig. 7. Use of analog to digital converter for current monitoring.

shows the position of the microprocessor in the digital shunt regulator system. Since this paper is concerned primarily with the microprocessor's role in voltage regulation, other signals and interfaces pertaining to possible housekeeping duties are not included in the diagram.

The processor's job is simply to add or to remove solar array sections from the bus to keep the shunt current between its maximum and minimum limits.

Figure 9 is a more detailed block diagram of the system's control loop. Many power systems have the inherent property that their loads don't change very often. A communications satellite, for example, might experience a load change only a few times per hour as transmitters are switched or antenna positions changed. The regulator's controller is configured to take full advantage of this system attribute.

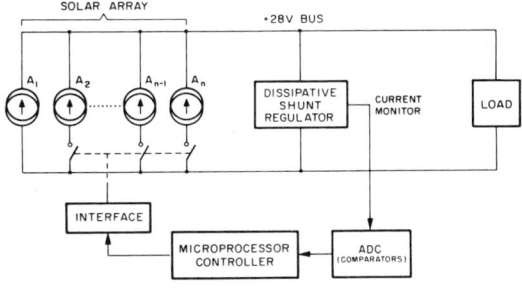

Fig. 8. Role of microprocessor in digital shunt regulator system.

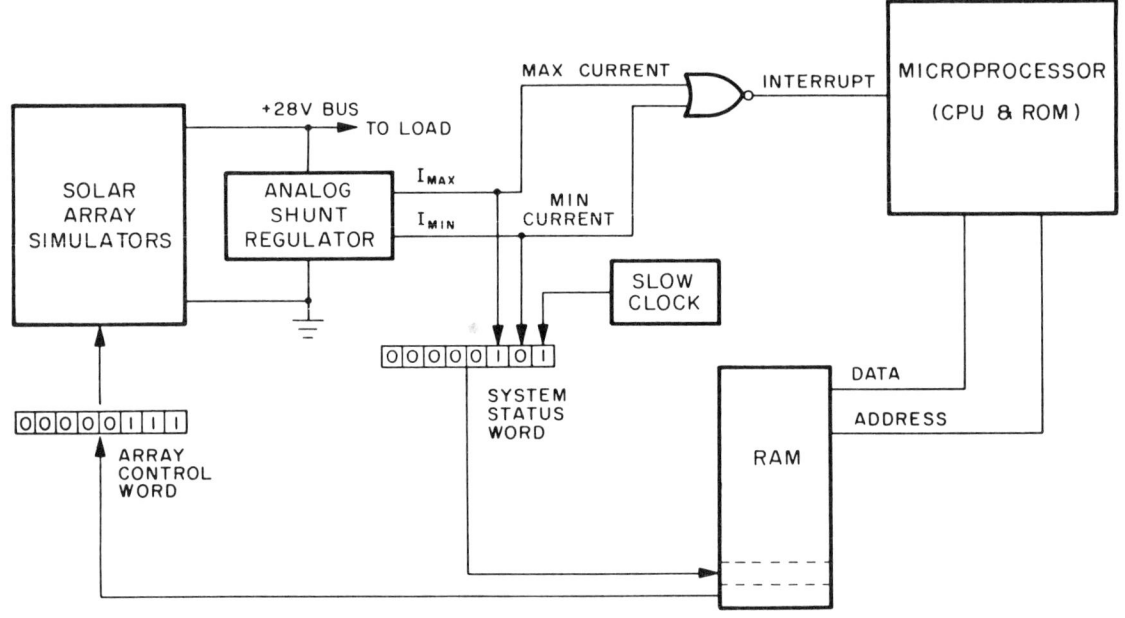

Fig. 9. System block diagram.

While the load is constant and the solar array is under constant illumination, the bus voltage should remain in regulation, any small changes being compensated by the self-contained analog shunt regulator. Under these conditions the microprocessor is free to do housekeeping tasks and other routine management jobs.

When a change does occur, however, and the shunt current level trips either the I_{max} or I_{min} comparator, the microprocessor is interrupted from its background job. The interruption initiates software which reads the system status, determines whether array sections must be added or removed from the bus, changes the array control word to reflect the power demand, and then sends the new control word to the solar array switches. The processor then resumes its background job and waits for another interrupt.

For maximum speed and simplicity, only one section of the solar array is switched per interrupt. The microprocessor is so fast, however, that the power system may not completely respond to the addition or removal of a section by the time the software is ready to return to the background job. If it were permitted to return at its maximum speed, the background job could be interrupted again immediately because the system would not have time to fully respond to the initial correction—even if the single addition or removal was adequate to compensate for the initial load change. The controller would then overcorrect and the system could be unstable.

To eliminate this possibility, the switching of the array sections and the return of system control to the background job are synchronous with a slow clock. This term is used to distinguish it, running at kilohertz, from the processor's clock, which runs at megahertz. The state of the slow clock is a part of the system status word that is read by the controller. Figure 10 shows timing relations with respect to the slow clock. The degree of size reduction of power components realized by the adoption of a digital regulator is a function of the speed of the slow clock. The maximum frequency of the slow clock is related to the bandwidth and transient response characteristics of the analog regulator. This frequency may be further restricted by the microprocessor if the software is slow. In

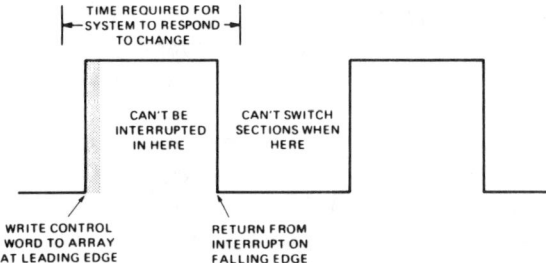

Fig. 10. Slow clock timing diagram assuring stability.

that case the microprocessor would limit the overall response and physical size reduction of the regulator. These considerations are discussed in greater detail in the next section.

A. Frequency Limitations

There are several factors which determine the characteristics of the slow clock. It is desirable to have the clock run as fast as possible, since the amount of capacitance required at the output of the shunt regulator is inversely proportional to the rate at which solar array sections can be added to the bus. The capacitor is required for energy storage during transient loads.

Assume there is a step increase in load current. The shunt regulator responds by reducing its shunt current until I_{min} is reached. Now no more current is available from the solar array and the capacitor must supply the balance of the load current until the microprocessor switches another array section onto the bus. The longer the processor takes to do this, the larger the capacitor must be to keep the bus voltage within specifications. A similar situation exists when there is a step reduction in load current. Then the capacitor must absorb the excess current until the processor removes a section from the bus.

The absolute maximum frequency which can be run is therefore determined by the time it takes for the microprocessor to respond to either an I_{max} or I_{min}, whichever takes more time.

The absolute maximum frequency may not be suitable for the system, however. The transient limits specified for the bus voltage and the absolute maximum frequency will together determine the minimum capacitance required on the regulator output. The capacitance will in turn influence the transient response characteristics of the shunt regulator. The frequency of the slow clock may have to be decreased to give the regulator time to respond to changes caused by the switching of array sections. Transients in the response must be given time to decay sufficiently before the processor is permitted to evaluate system status. Also, such over all system considerations as electromagnetic inference (EMI) or special synchronization requirements may influence the characteristics of the slow clock.

The software for the development system described in this paper requires 46 *processor clock cycles* to execute one switching action and return to a background job after an interrupt has occurred. At a 1 MHz rate this means that the slow clock can run at 21 kHz absolute maximum. A frequency of 10 kHz, however, was selected as a reasonable baseline for comparison with the performance of other system configurations.

For completeness it should be mentioned that the slow clock need not be operated at 50 percent duty factor as it is in this system, nor does it have to be periodic. A retriggerable single-shot could perform the same function by inhibiting the interrupt during the transient settling time.

In addition, the timing need not be derived from an independent source as it is done here. It may be more desirable in certain cases to derive a slow clock from the microprocessor's fast clock by means of a software-programmable frequency divider.

B. Hardware Implementation

Selection of the hardware used in the model regulator was influenced not only by the immediate availability of the individual parts but also by the availability of the equipment and facilities required to support them. The design, therefore, is not the most desirable. It should be mentioned, however, that the Rockwell 6502 central processing unit (CPU) was the first choice for the heart of the controller since its advanced architecture permits instruction execution in a minimum number of clock cycles and its zero page addressing feature is extremely useful in high-speed data transfer. Both these features permitted maximum design flexibility and system performance.

Memory consisted of two 2114 1K × 4 static random-access memory (RAM) chips and one 2708 1K × 8 UVPROM. Decoding was done with 74138's and the slow clock was made with a 555 timer. Interfacing was accomplished with a 74367 tristate bus driver and a 74174 latch. Miscellaneous NAND/NOR/NOT functions called for a 7400, 7402, and a 7404.

Although the parts count is small, it is much larger than required for this system. The circuit was purposely overdesigned in anticipation of facilitating system expansion for future development work.

C. Interfacing

The interfacing philosophy is very simple and straightforward. The address lines are decoded such that the lowest eight words of RAM are stolen for interfacing purposes. The addresses that were formerly in RAM are now external data ports. Figure 11 shows the allocation of the addresses. Figure 12 shows the decoding scheme and Fig. 13 gives the interface circuits.

Since this is only an elementary system only two of the eight available stolen addresses are utilized. If one wants to read the system status word, for instance, all that is required is to read the contents of address 0000. (Addresses here are written in hexadecimal.) Writing to that location is inhibited by the decoding hardware and has no effect. To switch the array sections all one must do is store the array control word at address 0003. Attempts to read from that location are inhibited and have no meaning.

It is easy to see that it would be a simple matter to control or monitor many functions with just these eight locations, since each word contains eight bits of information. It is also a trivial matter to extend the interfacing addresses from eight to a higher power of 2 just by modifying the decoding circuit.

The interface to the processor's interrupt request line ($\overline{IRQ}$) is through a single NOR gate. Thus if the shunt regulator demands attention for either too much or too little current, the CPU is alerted.

D. Software

Figure 14 shows the entire program for the processor. The program as shown is written in 6500 mnemonics and coded for input to a cross-assembler. The flowchart for the operations is given in Fig. 15.

The program was written to execute the array switching operation in the fewest number of clock cycles to achieve a high value for the absolute maximum slow clock frequency discussed in Section 7A. Extensive use is made of the zero page addressing feature of the 6502.

The background job is just a dummy do-nothing routine which idles the processor when no demands are being made on the system.

8.0 Complete System

This section summarizes the operation of the complete DSR system. When the power is first turned on, the dissipative analog shunt comes on immediately. As the bus voltage slowly builds up,

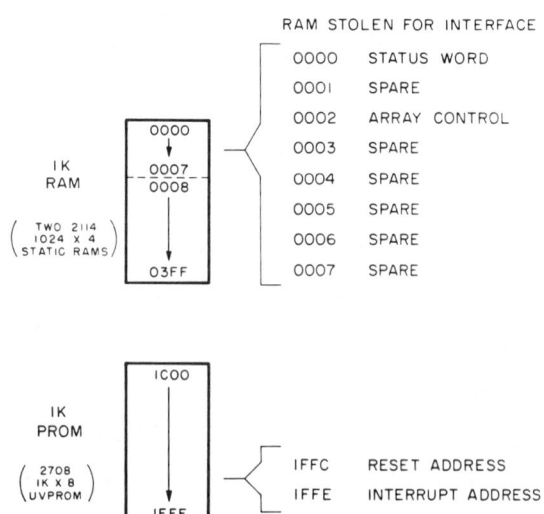

Fig. 11. Address allocation.

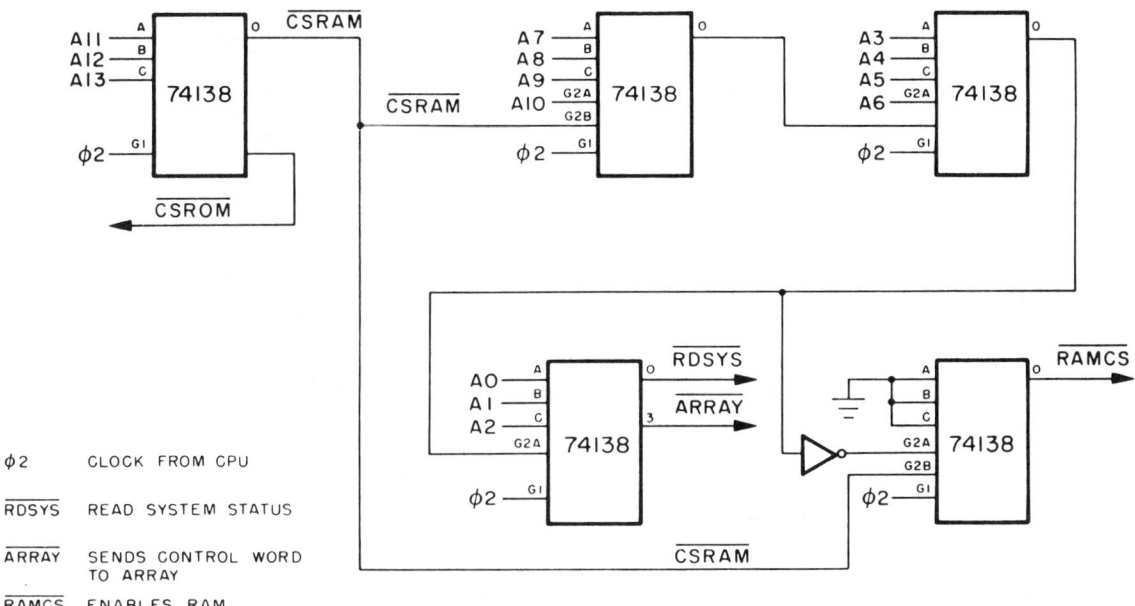

Fig. 12. Decoding scheme.

no shunt current flows until the bus voltage crosses 28 V, after which the bus voltage is maintained by controlling the current in the shunt such that the sum of the load current and the shunt current is equal to the current of the solar cell array section at 28 V.

If the load current increases, the shunt current decreases and vice versa. If the load current continues to increase such that the shunt current reduces below the I_{min} value, then the I_{min} comparator output activates the microprocessor to switch one more section onto the bus. Now the shunt current rises to a level between I_{max} and I_{min}. Again assume that the load current increases further such that the shunt current reduces below I_{min}. Then the processor will switch on one more section just as before. This process continues until all sections are switched on. Now the load current can increase no more as the system is designed only to this maximum value of load current.

Now say the load current decreases. Then the shunt current increases to maintain the bus voltage. But assume that the load current decreases such that the shunt current increases above the

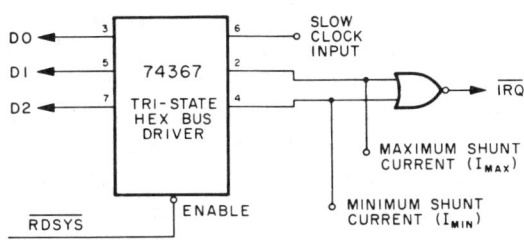

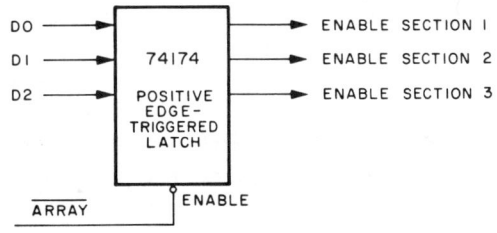

Fig. 13. Interface circuits.

```
                TITLE "DIGITAL SHUNT REGULATOR"                          BEQ   MIN           ;IF NOT GO TO MIN ROUTINE
                                                                         LDA   SETREG        ;REMOVE ONE SECTION OF ARRAY
; THIS SECTION DEFINES RAM LOCATIONS                                     ASLA
        ASECT $0000                                                      ORA   #$01          ;SHIFT IN A "1"
SYSTAT:         BLOCK 1         ;CLK, MIN, MAX                           JMP   CLKCHK        ;NOW WAIT FOR PROPER TIME TO WRITE

        ASECT $0003                                             MIN:                         ;NEED MORE CURRENT--ADD ONE SECTION
ENABLE:         BLOCK 1         ;DSR5, DSR4, DSR3                        LDA   SETREG
                                                                         LSRA                ;SHIFT IN A ZERO
        ASECT $0010
SETREG:         BLOCK 1         ;ARRAY CONTROL WORD             CLKCHK:                      ;CHECKS FOR POSITIVE EDGE OF CLOCK
                                                                         LDX   SYSTAT        ;WAIT FOR CLOCK LOW
        ASECT $0050                                                      CPX   #$03
STACKTOP:       BLOCK 20        ; RESERVED FOR STACK                     BPL   CLKCHK

                                                                WRITERDY:                    ;WRITE WHEN CLOCK GOES HIGH
; THIS SECTION DEFINES ROM LOCATIONS                                     LDX   SYSTAT
        ASECT $1FFC                                                      CPX   #$03
        BKWORD RESET            ;PROGRAM GOES HERE ON RESET              BMI   WRITERDY      ;LOOP IF CLOCK STILL LOW
                                                                         AND   #$07          ;KEEP ONLY 3 LOWEST BITS
        ASECT $1FFE                                                      STA   ENABLE        ;WRITE CONTROL WORD TO ARRAY
        BKWORD ATTN             ;PROGRAM GOES HERE ON INTERRUPT          STA   SETREG        ;STORE CONTROL WORD IN RAM

        PSECT                                                   CLKFALL:                     ;CHECKS FOR NEGATIVE EDGE OF CLOCK
; INITIALIZATION ROUTINE                                                 LDX   SYSTAT
                                                                         CPX   #$03
RESET:  LDA   #$07              ;TURN OFF CURRENT SOURCES                BPL   CLKFALL       ;LOOP IF STILL HIGH
        STA   ENABLE                                                     CLI                 ;IF NEGATIVE EDGE, CLEAR INTERRUPTS AND RETUR
        STA   SETREG            ;INITIALIZE CONTROL REGISTER             RTI
        CLD                     ;CLEAR DECIMAL MODE
        LDX   #$50              ;SET STACK POINTER              ; BACKGROUND JOB
        TXS
        JMP   BACKGROUND        ;GO TO BACKGROUND JOB TASKS     BACKGROUND: CLI              ;ENABLE THE INTERRUPTS
                                                                LOOP1:   LDY   #$20          ;ROUTINE PRINTS CHARACTERS
                                                                LOOP2:   CPY   #$5F          ;TO SIMULATE OTHER POWER
ATTN:                           ;PROGRAM GOES HERE ON INTERRUPT          BEQ   LOOP1         ;MANAGEMENT TASKS
        LDA   SYSTAT            ;LOAD ACCUMULATOR WITH SYSTEM STATUS     STY   $#411
        AND   #$01              ;CHECK TO SEE IF MAX BIT IS SET          INY
                                                                         LDX   #$00
                                                                WAIT:    INX
                                                                         CPX   #$FF
                                                                         BNE   WAIT
                                                                         JMP   LOOP2
```

Fig. 14. Program.

I_{max} value. Then the I_{max} comparator output is used by the microprocessor to remove one section from the system. The shunt current will then decrease. Assume now that the load current decreases again such that the shunt current increases to give an I_{max} output. One more section will be switched off the bus. If the load current is decreased to zero this process continues until all sections are switched off and disconnected from the bus. As one section is connected permanently without a switch, one section will be on all the time. But as the shunt is designed for one section full power, the bus voltage is maintained even if the load is completely disconnected.

Thus the DSR maintains the bus voltage at fixed level from no load to full load.

9.0 Experimental Results

A model system of a microprocessor controlled digital shunt regulator was built as described earlier in this paper. Data taken from that setup is presented here to illustrate the system's performance under several interesting conditions.

Table I shows the dc steady-state values for system parameters under several different loads. The I_{max} threshold was set for 400 mA and the I_{min} value was set for 45 mA. These values were selected according to the measured I-V characteristics of the solar array simulators (Fig. 4). The levels must be chosen to be certain there is no possibility of overlap from I_{min} to I_{max} when a section is switched onto the bus. Some margin is included to allow for the presence of noise on the sense lines. The dc regulation is ± 30 mV from no load to full load. One can also see from the table that a section is switched when a change in load current causes the shunt current to cross either an I_{max} or I_{min} threshold. The limit of regulation is about 1.15 A, when the shunt current drops below the I_{min} value and there are no more sections left to be switched onto the bus. A little more load current causes the output voltage to fall with increasing load.

The remaining data is for dynamic conditions. Figure 16(A) shows the response of the regulator

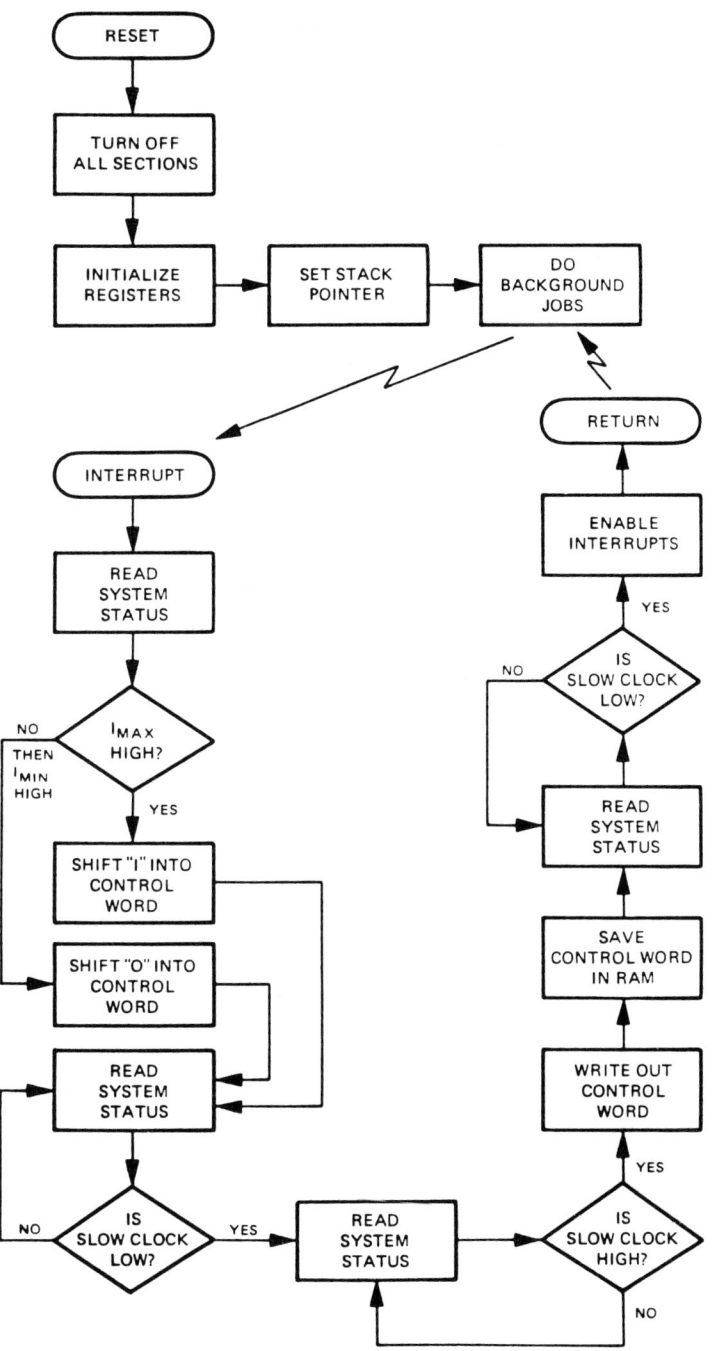

Fig. 15. Flowchart.

Table 1. System Status at Various Loads.

Load Current (mA)	Shunt Current (mA)	Status of Switched Array Sections			Bus Voltage (volts)
		S1	S2	S3	
0	290	OFF	OFF	OFF	28.02
200	90	OFF	OFF	OFF	27.98
300	310	ON	OFF	OFF	28.02
400	210	ON	OFF	OFF	28.00
500	110	ON	OFF	OFF	27.98
600	320	ON	ON	OFF	28.03
700	220	ON	ON	OFF	28.00
800	120	ON	ON	OFF	27.98
900	280	ON	ON	ON	28.03
1000	180	ON	ON	ON	28.01
1100	80	ON	ON	ON	27.98
1150	30	ON	ON	ON	27.97

to a dynamic load under the condition that the load does not cause the shunt current to cross a threshold. The top trace is the ac portion of the bus voltage. The center trace is the load current at 200 mA/div and the bottom line marks zero load current. The load current experiences a 250 mA step at a dc load of 600 mA. The output voltage falls by about 60 mV. No switching of array sections has occurred. The bus voltage drop is just the dc regulation of the shunt regulator. The small slope in the voltage waveform is due to the ac coupling of the oscilloscope preamplifier being used at the low frequency 20 Hz pulse repetition frequency (PRF) of the dynamic load.

Figure 16(B) is the same as above except that the dc load level is at 900 mA (current is 500 mA/div). The voltage waveform is virtually identical to that in the previous case since the shunt current is nearly the same. The extra load current is being supplied by one more array section which the processor has added to the bus.

Figure 16(C) is the response to a dynamic load when the load change causes an extra section to be switched. Bus voltage is at 200 mV/div. The current is at 500 mA/div. The dc level is 700 mA and the step change is 400 mA. At the leading edge of the step the bus voltage falls until one section is added to the bus. Then the bus voltage rises to resume its proper regulated level. When the load transient is removed there is temporarily too much current being supplied (from the extra section) and this causes the bus voltage to rise. This excess current is soon detected by the processor, however, the section is switched off, and the regulated bus voltage is restored.

Figure 16(D) is under conditions same as above but the leading edge is expanded in time and the slow clock is shown at the bottom to illustrate the timing relationships. Note that the array section is not switched (evidenced by the rising bus voltage) until the slow clock goes high as prescribed by the software.

Figure 16(E) is identical to Fig. 16(D) except that the load transient has occurred at a different place in the period of the slow clock. The processor was interrupted almost immediately at the leading edge of the step, but recall that it takes 46 μs to execute a switching action. It hasn't enough time to switch the array in the present slow clock period and therefore has to wait until the next time the slow clock goes high. Thus the output voltage is allowed to fall much farther than in the previous case. The fact that the processor may have to wait one entire slow clock period *plus* the 46 μs execution time should be taken into account when computing worst case transient bus voltages.

Figure 16(F) illustrates what happens when the load transient is enough to warrant two sections to

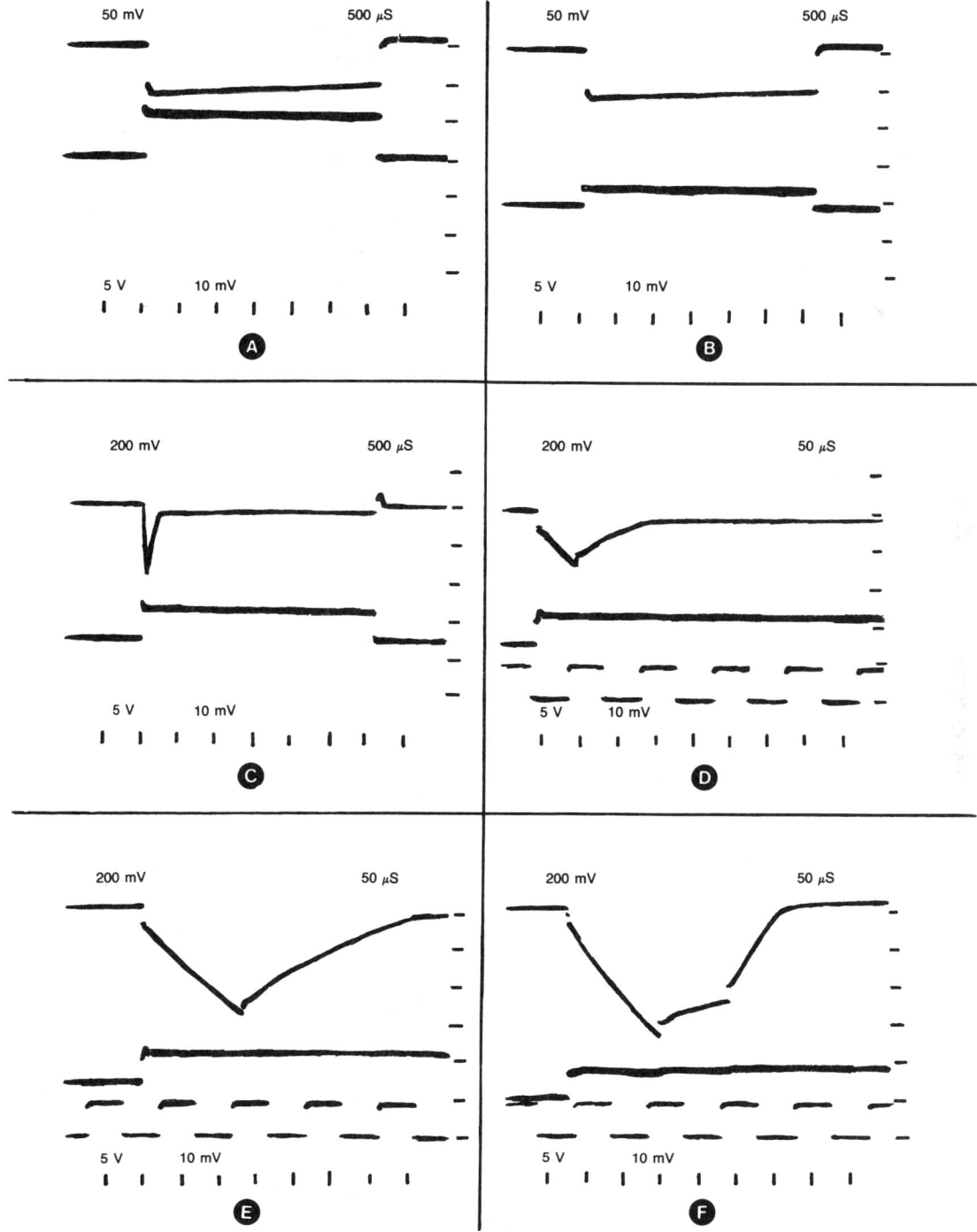

Fig. 16. Waveforms under dynamic loading.

be switched onto the bus. Current is at 500 mA/div, dc level is 500 mA and, the step is 400 mA. The slow clock is again at the bottom. One can clearly see the two array sections being added to the bus by noting the changes in slope of the bus voltage.

10.0 Extensions

Figure 17 shows the redundant systems of a microprocessor controlled DSR. Though two microprocessors are used, because of cross connection, the reliability has been further enhanced. In addition, this approach is expected to result in an integrated system for controlling DSR, battery management, and for controlling CR/DR as shown in Fig. 18.

The same microprocessor can also be used to monitor various housekeeping parameters in a spacecraft power system or in a ground based power system.

11.0 Conclusions

For solar power systems a digital shunt regulator is superior to other types of shunt regulators, and the use of a microprocessor for the control of the digital shunt regulator results in improved system performance. System flexibility is

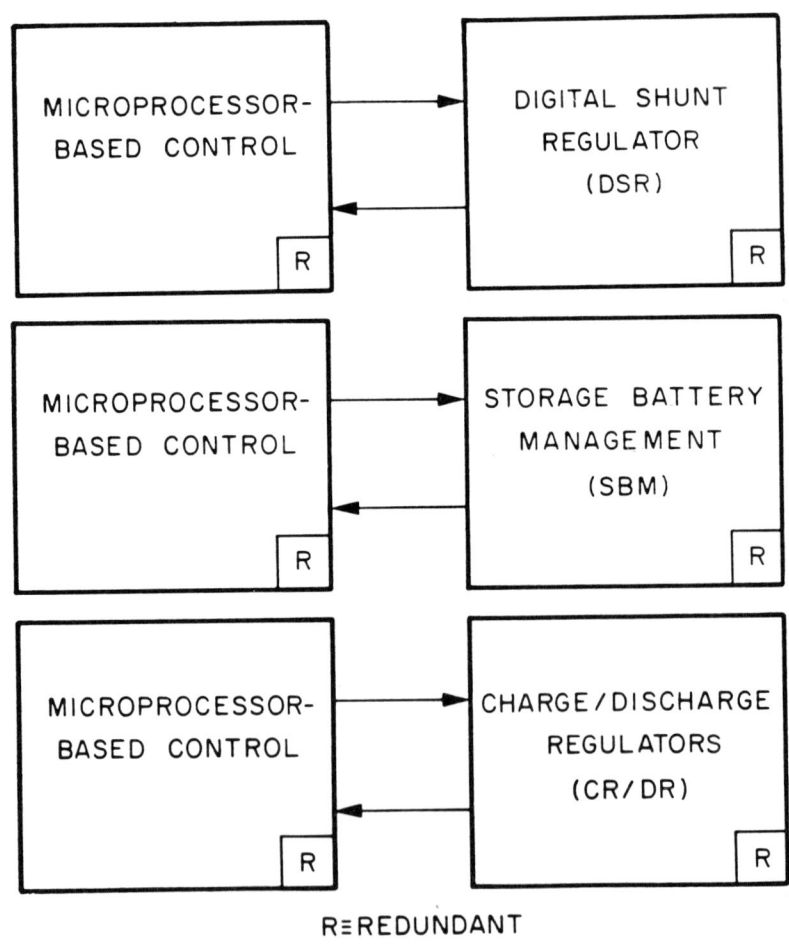

Fig. 17. Redundant systems.

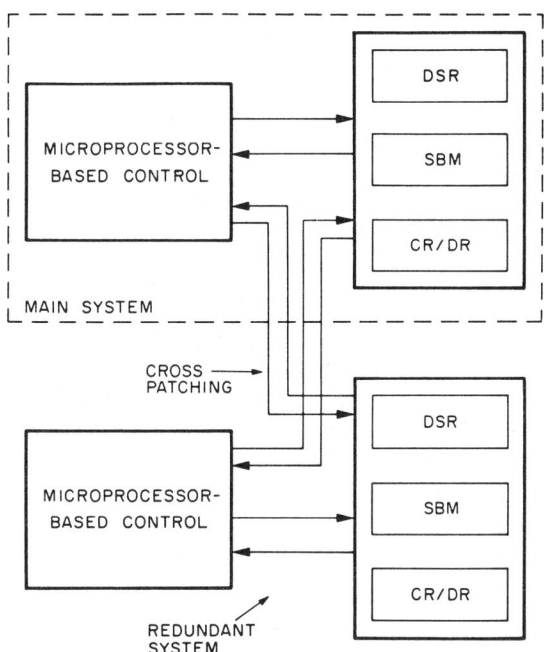

Fig. 18. An integrated system for digital shunt regulator, charge regulator/discharge regulator, and battery.

the chief advantage of a microprocessor based system.

The advantages arise from the ability to replace hardware with software, permitting decisions related to design parameters to be made at a later stage in the project. For example, if the system requires a modification, the change can be implemented by changing the software only, or at worst, software and minimal hardware. Such modifications are simple and less time consuming to implement than previous solutions which involve major hardware design changes. Thus the system capability is enhanced, flexibility is increased, and the design is faster and less expensive than the conventional approach. Moreover, the system can be modified in real time in response to natural component degradations or to anomalies.

Since the microprocessor need not be dedicated to the regulator, it can simultaneously be used for battery management and for charge regulator/discharge regulator control. This feature reduces overall component count, simplifies assembly and testing of the unit, and results in significant time saving. Because the overall system component count can be reduced the reliability can be increased. Implementation of a redundant system is easily done to further enhance the high reliability of the power system.

Acknowledgments

The authors wish to acknowledge the assistance of the Caltech Computer Science Department, whose facilities were used during development and testing of the microprocessor controller.

References

1. P.R.K. Chetty, "Spacecraft power systems—some new techniques for performance improvement," Ph.D. Thesis, Indian Institute of Science, Bangalore, India, 1978.
2. M.S. Inamure et al., "Microprocessor controlled battery protection system" IECEC 1975 Rec., pp. 1307-1717.
3. C. Gayet, "Battery management using microprocessors," ESA-SP-126, pp. 251-262, 1977.

MULTIPHASE OPERATION OF SELF-OSCILLATING SWITCHING REGULATOR

Introduction

With the continual improvement in the degree of sophistication, new demands are being made on the power conditioning systems such as switching regulators and dc-dc converters. Switching regulators have continued to gain popularity despite their tendency to generate EMI, because they exhibit or offer high efficiency over a wide range of input voltage. Low power loss also eliminates the need for large heat sinks and reduces the cost of the unit. In addition most of the significant developments in monolithic IC regulators have further enhanced the use of the switching regulators. Because of power handling limitations of semiconductor devices, such elements are quite often used in parallel in switching regulators to meet the high power requirements. In such a situation, the power is shared in phase between the power handling devices, since all of them switch simultaneously. This simultaneous switching action creates problems in filtering and electromagnetic screening. Multiphase operation is employed primarily to minimize these problems.

The principle of multiphase operation of power conditioners of driven type has been already reported 1, 2, 3. Reference [2] has dealt with a driven type multiphase PWM shunt regulator whereas reference [3] dealt with a driven type two phase 100 watt PWM boost regulator. But there is no report of multiphase operation of the self-oscillating power conditioners. Self-oscillating power conditioners are preferred for their reduced complexity, size and cost compared to driven type switching power conditioners. The multiphase operation of self-oscillating power conditioner (switching regulator) is described below.

Theory of Operation

The principle of multiphase operation in self-oscillating power conditioners is implemented for PWM buck type regulator and similar implementation is possible for other system of power conditioners like PWM shunt regulator, PWM boost regulator, etc. The proposed PWM buck type regulator has been designed to work either on regulated bus or on an unregulated bus. Figure 1 shows a two phase version of the buck type regulator which can be extended to more number of phases as shown in dotted line. It comprises a ratio network to reduce the output voltage, a reference voltage source (Vref) and an error amplifier IC1) to provide the PWM control signal. This signal is given to switch S1 and also to the phaseshift network. The phase shifted control signal is given to the switch S2. Thus the series switches S1 and S2 will be ON for a fixed duration and OFF for a fixed duration, but not simultaneously. There is a finite time lag between the two switches S1 and S2, decided by the phase shift network which can be controlled but the duty cycle is maintained. If the self-oscillating frequency is f(= 1/T), then the time delay adjusted is T/2 because this regulator is of two phase version. If this regulator is of n-phase version, then the phase shift or time delay from each other switch (S1,S2,.....Sn) will be adjusted to T/n.

The self-oscillating frequency of a regulator working on a regulated bus is almost constant, while on unregulated bus where the bus voltage varies over a range, the frequency varies over a small range. If a self-oscillating switching regulator of the type, what W.A. Peterson described [4], is employed then the operating frequency is stabilized

---

Extracted from Ph.D Thesis of P.R.K. Chetty, entitled *"Spacecraft Power Systems—Some New Techniques for Peformance Improvement,"* Indian Institute of Science (I.I.Sc), India, 1978. Portions are reprinted with permission from the *Proceedings of the third ESTEC Spacecraft Power Conditioning Seminar,* ESA-SP 126, pp. 149-153, Sept 21-23, 1977.

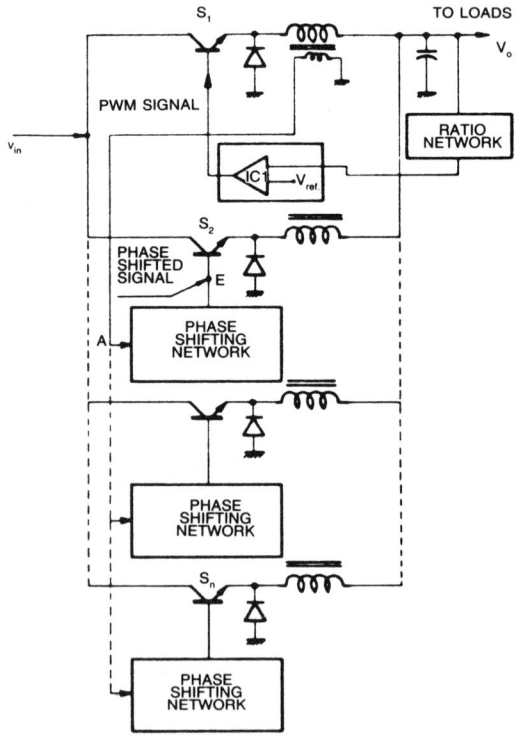

Fig. 1. Two phase buck regulator. Extension to n-phase is shown in dotted line.

even if the input voltage varies. In any case, the phase shift or delay is adjusted with respect to higher frequency or lower period. The effect of this mode of operation is that the current is pumped sequentially into an output load at double the signal frequency. If there are n similar units, then the effective frequency is n times the signal frequency.

Description of Phase-Shift Network

Figure 2 shows the schematic of the phase shift network. This network consists of two monoshots and two exclusive-OR gates. The PWM control signal shown in Fig. 2A triggers both monoshots. Monoshot-1 is positive edge triggered and its output is shown in Fig. 2B. This is exclusive-ORed with the input PWM control signal and its output is shown in Fig. 2D. Monoshot-2 is negative edge triggered and its output is shown in Fig. 2C. This is exclusive ORed with the output of the previous exclusive-OR gate output and is shown in Fig. 2E. Thus Fig. 2E shows the phase shifted PWM control signal whose duty cycle and frequency are the same as the original PWM control signal. The output pulse width of monoshots are adjusted to a period equal to T/2.

Any type of monoshots can be employed here, i.e., using a single NAND gate to two NAND gates or monoshot IC(SN74121) or discrete version. Similarly the exclusive-OR gate can be SN7486 or discrete version [5].

Design

The following specifications are considered for the design of two phase self-oscillating switching regulator.

Input voltage (Vin)	=	20 to 30 volts
Output voltage	=	12 volts
Output current(Io)	=	1 amp
No. of phases	=	2
Maximum output current	=	1.1 amp
Output regulation	=	± 1%
Output ripple (Vpp)	=	< 100 mV at load current
Frequency of operation	=	4-8 kHz
Effective frequency of operation	=	8-16 kHz

The control block shown in Fig. 3 is used to sense the output voltage and to provide a pulse width modulated signal which when used to control the transistor switches in the power stage, results in a regulated output voltage (V_o). An IC regulator is used in the design of this block as this is small, efficient, can switch very fast. This IC has also a built-in temperature compensated voltage reference source (V_{ref}). Scaled down output voltage V_o and built-in V_{ref} serve as inputs to IC regulator error amplifier. The output of IC regulator is a PWM waveform. The frequency of the PWM waveform varies as a function of input voltage, output voltage, load current, ripple requirement, hysteresis of the IC regulator error amplifier, the values of the output filter network. Small positive feedback is applied by employing R7 and

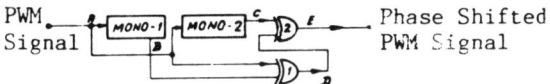

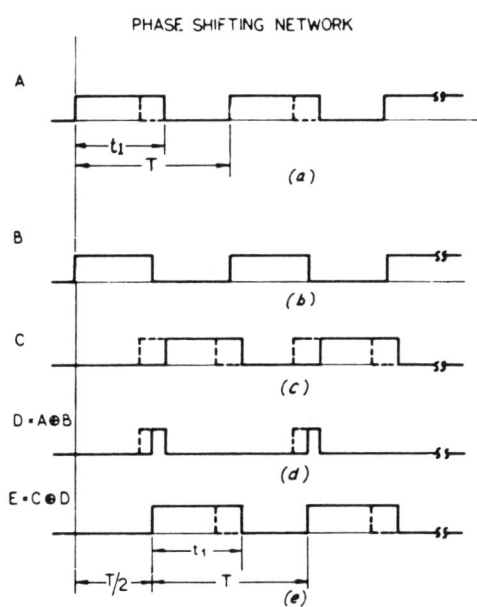

Fig. 2. Phase shifting network and its waveform at different points.

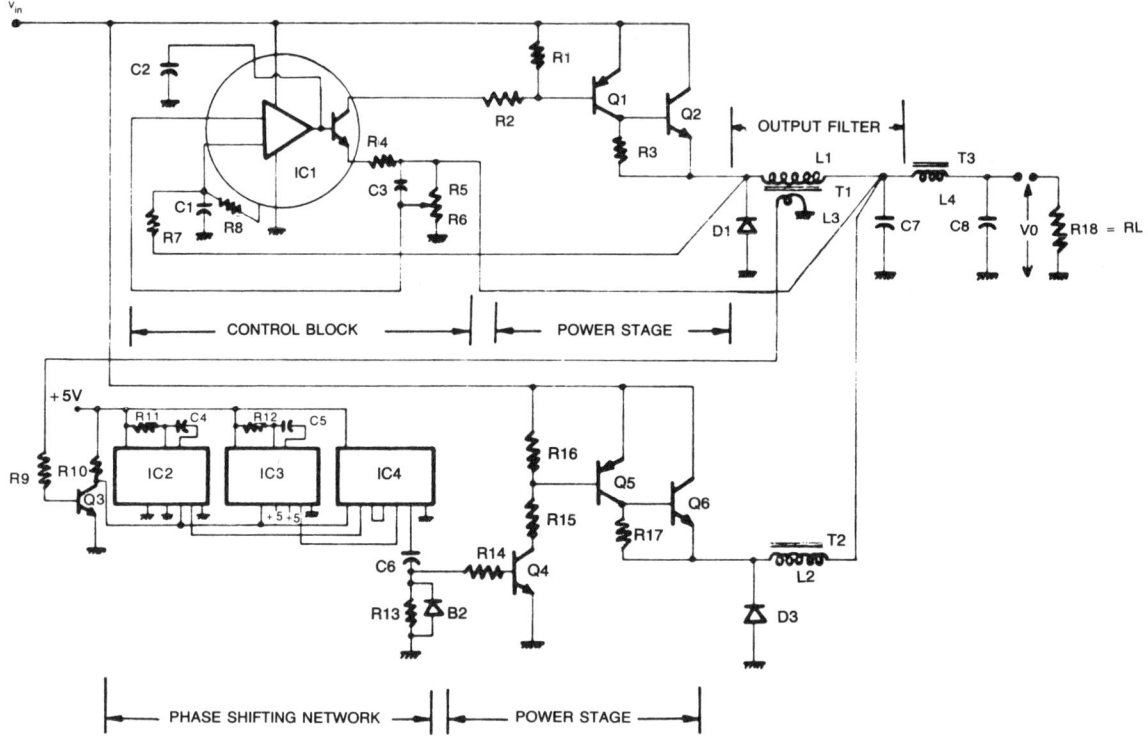

Fig. 3. Circuit diagram of two phase self-oscillating Switching Regulator.

R8 from the output for positive oscillations. The frequency is chosen depending upon the frequency response of the available components, mainly the power transistors, switching diodes, etc., and to keep the weights of inductor and capacitor to a minimum. The value of C1 is selected experimentally for proper frequency compensation. The value of C2 and C3 are selected experimentally for keep-

Table 1. Performance of Two Phase Self-Oscillating Switching Regulator.

Vin (Volts)	Iin (mA)	Vo (Volts)	Io mA)	n (%)
18	800	11.93	1000	81.0
20	740	11.95	1000	80.7
22	685	12.01	1000	79.7
24	625	12.09	1000	78.9
26	585	12.09	1000	77.8
28	530	11.95	1000	78.8
30	500	11.93	1000	77.9

Self-oscillating frequency changed from about = 4 to 8.5 kHz
Regulation (Vin = 18V – 30V) = < +/– 1%
Ripple (Vin = 18V – 30V) = < 60 mV

Table 2. Comparison of Conventional Approach and Multiphase Approach for 4 Phase Operation.

Components required for

Conventional Approach	Multiphase Approach
1 Regulator IC	1 Error amplifier
8 EX-OR gates	2 Voltage references
6 Monoshots	2 J-K Flipflops
4 Drivers	8 NAND gates
	4 Sawtooth generators
	4 Comparators
	4 Drivers

ing the output ripple to a minimum. The values of L(L1,L2) and C7 are calculated using the formulae given below:

$$L = \frac{(V_{in} - V_o) V_o}{2 \cdot f \cdot (I_{omax} - I_o) V_{in}}$$

$$C7 = \frac{(V_{in} - V_o) V_o}{8 \cdot V_{in} \cdot L \cdot f \cdot f \cdot (V_{pp} - V_h)}$$

where V_h is a hysteresis of the IC regulator error amplifier.

Experimental Results

Figure 3 shows the detailed circuit diagram of the two phase self-oscillating switching regulator. This regulator was built and tests were carried out to evaluate its performance. Table 1 gives the performance of this regulator as observed under various input voltages. It is easy to see that the performance closely follows the specifications.

The principle implemented here for buck type self-oscillating regulator, can also be implemented for boost and shunt regulators. Implementation of this principle to the driven type switching regulators reduces the parts count which is clear from Table 2 of comparison of conventional approach and multiphase approach for a four phase operation.

References

1. A. Bazin, "Etude du circuit de controle multiphase de la commande de n etages de puissance en parallel," Contract ESTEC No. 830/69/HP, 1970.
2. D.M. Sulliven, "The multiphase PWM shunt," ESRO-SP-84, pp. 135-147, 1972.
3. A. Carraro, "Multiphase PWM boost regulator," Proceedings of Spacecraft Power Conditioning Seminar, 1977, ESA-SP-126, pp. 137-147.
4. W.A. Peterson, "A frequency stabilized free running dc to dc converter circuit employing pulse width modulated control regulation," IEEE Power Electronics Specialists Conference 1976, pp. 200-205.
5. P.R.K. Chetty, "Exclusive-OR circuit handles wide range of input levels without power supply," *Electronic Design,* Vol. 22, ED-4, pp. 78, 1974.

DC-DC CONVERTER MAINTAINS HIGH EFFICENCY

Introduction

Dc-dc converters are widely employed in applications where there is a need for a particular voltage to operate an electrical or electronic system or equipment than the source voltage. Thus the dc-dc converters are used for stepping-up, stepping-down and for electrical isolation. These converters, when supplying constant power load, such as a PWM switching regulator, exhibit a lower efficiency at high input voltages than at low input voltages. The PWM switching regulator draws constant power over a wide range of input voltages. A wide range of input voltages is normally expected when the spacecraft passes from sunlit portion of the orbit to eclipse, or vice-versa. Under normal circumstances, the converter has to supply a constant power to the load, even when the input voltage undergoes large variation because the power load does not care for the input voltage variation and it takes always constant power.

The approach to the design of dc-dc converter presented here overcomes the above mentioned limitation. An analysis is carried out to show that the efficiency of the conventional converter decreases as the input voltage increases. The conditions are derived for maximum efficiency over a range of input voltages. A control circuit is designed to satisfy these conditions and the results of the practical implementation are presented.

Converter Operation

Figure 1 shows a simple dc-dc converter which is a free running multivibrator that utilizes the saturating effect of a transformer core and the switching properties of the transistors to generate square waves. The square wave can be transformed-up or -down and rectified to produce pure dc. In practice little filtering is required. As the converter uses transistors and square loop magnetic cores, it operates successfully at higher frequencies, and hence is lighter and more efficient.

Analysis

Consider the dc-dc converter shown in Fig. 1. Let Po be the output power, Pin the input power, and n, the efficiency of the converter. Then

$$Pin = Po/n \quad (1)$$

The collector or primary current, Ic, neglecting Vce sat of the transistor to simplify the calculation, is given by the relation

$$Ic = Pin/Vg = Po/[n \cdot Vg] \quad (2)$$

where Vg is the input voltage. Hence the base current should be

$$Ib = Ic/[hFE\ min] = Po/[n \cdot Vg \cdot hFE\ min] \quad (3)$$

where hFE min is the minimum shortcircuit gain at Ic. The base drive is usually made equal to twice the maximum base to emitter voltage of the transistor plus starting diode voltage drop, Vd, to reduce the effect of differences in Vbe between the two transistors. Therefore the feedback voltage, Vf, is equal to [2Vbe max + Vd]. The base resistor Rb is chosen to drop approximately Vf/2 and hence is given by

$$Rb = [n \cdot Vg \cdot Vf \cdot hFE\ min]/[2 \cdot Po] \quad (4)$$

Rewriting

$$n = [2 \cdot Po \cdot Rb]/[Vg \cdot Vf \cdot hFE\ min] \quad (5)$$

or

$$n = [2 \cdot Po \cdot Rb]/[Vg \cdot Vg \cdot k \cdot hFEmin] \quad (6)$$

since Vf = k · Vg where k is the feedback turns ratio. From Eq (6), it is clear that the efficiency of the converter is inversely proportional to the square of the input voltage as the remaining parameters are approximately constant. Thus, if the converter is designed for an input voltage Vg1 and is used over a range of input voltages higher than Vg1, its efficiency decreases with the increase in input voltage.

To derive the condition for maximum efficiency, the first derivative of the Eq(5) is set equal to zero. Thus Eq(5) can be written as

$$n = c/[Vg \cdot Vf] \quad (7)$$

where c is equal to [2 · Po · Rb]/[hFE min] and

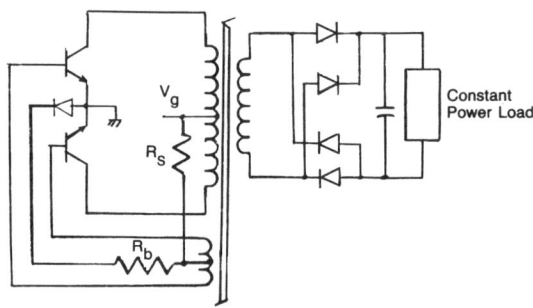

Fig. 1. Conventional dc-dc converter.

Extracted from Ph.D Thesis of P.R.K. Chetty, entitled "*Spacecraft Power Systems—Some New Techniques for Peformance Improvement,*" Indian Institute of Science (I.I.Sc), India, 1978. Portions are reprinted with permission from *ELECTRONICS,* Jan. 3, 1980. Copyright © 1980. McGraw-Hill Inc. All rights reserved.

is approximately constant for a particular circuit. Now maximizing the efficiency, n,

$$n = -\frac{c}{(Vg \cdot Vf \cdot Vf)}(dVf)$$

$$-\frac{c}{(Vf \cdot Vg \cdot Vg)}(dVg) = 0$$

Therefore

$$[dVg/Vg] = [dVf/Vf] \quad (8)$$

When the input voltage changes, say from Vg1 to Vg2, and the corresponding feedback voltage from Vf1 to Vf2, by integrating Eq(8) and applying the limits results in

$$[Vg2/Vg1] = [Vf1/Vf2] \quad (9)$$

This means that the feedback voltage has to be changed to correspond with the changes in input voltage such that it satisifies Eq(9). Thus, if the feedback voltage has to be programmed, it is necessary to have a number of tappings on the feedback winding. This is a complicated and non-practical solution. Again there is a minimum feedback voltage which is required for the converter to function well. Of the other terms in Eq(1), Po is constant and hFE is almost constant. Therefore Rb is the only parameter that may be considered for programming.

Using Eq(5) and maximizing for n with respect to Vg and Rb, and keeping Vf as a constant, results in

$$[dRb/Rb] = [(2 \cdot dVg)/Vg] \quad (10)$$

This means, for two specific values of Rb(Rb1 and Rb2) corresponding to Vg1 and Vg2,

$$[Rb2/Rb1] = [Vg2/Vg1]^2 \quad (11)$$

i.e., the resistor, Rb, is to be varied such that the ratio of minimum to maximum values is in direct proportion to the square of the ratio of minimum to maximum input voltages. Obviously the required variation in the value of Rb is non-linear. Thus the base drive has to be controlled as a function of input voltage per Eq(11) to achieve maximum efficiency over a wide range of input voltage, the implementation of which is dealt in the following section.

Practical Implementation

A control circuit is designed to satisfy Eq(11) and is shown in Fig. 2B. Three transistor (Q3,Q4,Q5) controller and two active base resistor networks drive Q1-Q2. As V_g increases, the voltage at point "b" increases. R2 and R3 are selected so that Vb is about 1 volt at Vin min, enabling Q3 to operate in active region. Because the collector of Q3 is biased from a reference (point a), the drive

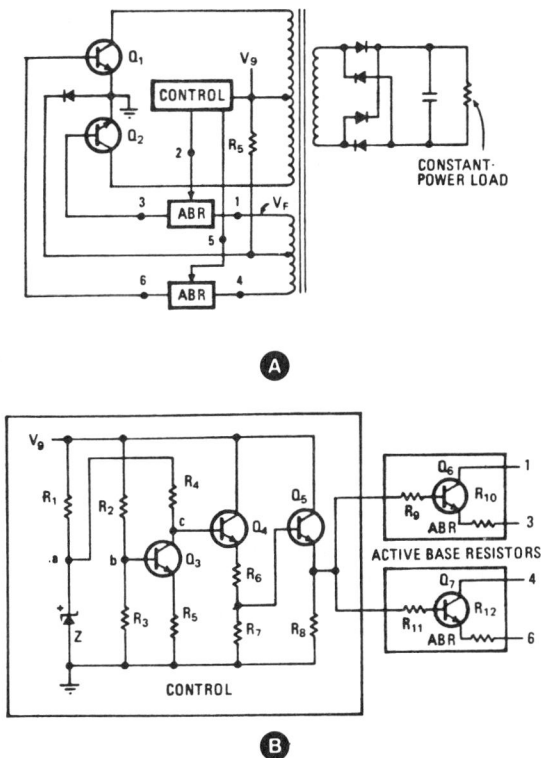

Fig. 2. New improved dc-dc converter. (A) Block schematic (B) Detailed circuit diagram of control circuit and active base resistors.

signal applied to Q4 varies as a function of the voltage applied to Q3's base. When the voltage at point b increases, the potential at point c decreases. As Q4 and Q5 are configured in emitter follower mode (with different gains), the switching transistors Q1-Q2 through Q6-Q7 are driven with less base current. As a result, the resistance between the points 1-3 (and 4-6) varies approximately as square of the input voltage to satisfy Eq(11).

Only one operating variable is be determined empirically, the voltage at the base of Q5, Vbq5. To do this, the circuit is broken at this point, an external variable voltage source is connected. Vin is set to its minimum expected value. The variable voltage source is then set just to saturate Q1 and Q2 for a constant Po, and its value (Vb1) is noted. The procedure is repeated to find Vb2 for Vin max. Now, R6 and R7 are determined such that Veq5 equals Vb1 at Vin min and is equal to Vb2 at Vin max.

Experimental Results

Figure 3 shows the efficiency Vs input voltage of the conventional and improved dc-dc converter. The improved dc-dc converter exhibited almost constant efficiency over an input voltage range of 20 to 40 volts, while supplying a constant power of 8 watts. The small deviation in efficiency from constant efficiency is because of the control unit (active base drive) power dissipation, etc.

Proportional Base Drive

From the above discussions, it is clear that the efficiency of the system or transistor can be improved if it is driven just enough to keep it in saturation, and not into deep saturation, although the collector current varies over a range. Also this type of transistor operation reduces the transistor's (saturation) turn-off time as there are less excess charges to remove. Indirectly turn-off losses are reduced plus the transistors can be operated at higher frequencies. Thus proportional drive has the advantages of high efficiency, higher frequency operation, etc. The approach employed in the previous sections does not employ any transformer.

However, recent proportional drives employ transformer and the whole operation of proportional drive depends on transformer coupling. In this type of proportional base drive, which is shown in Fig. 4, the transformer has three windings, i.e., trigger winding, base winding and collector or emitter winding. Through the trigger winding a narrow drive pulse is applied which starts the transistor conduction. As the collector current builds up, due to coupling, the base winding develops base drive proportional to collector current. This coupling and the transformer action continues as long as the col-

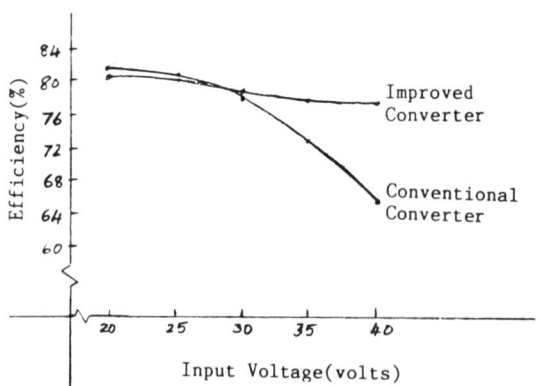

Fig. 3. Efficiency vs input voltage of conventional and new improved dc-dc converters.

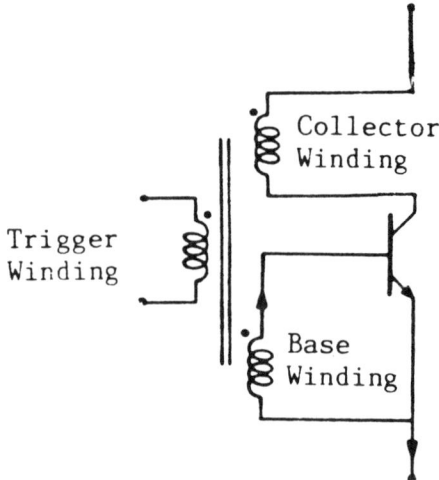

Fig. 4. Proportional base drive using transformer.

lector current varies (increases only or decreases only). The transistor is turned-off by applying a reverse base current through the trigger winding.

Thus, the transformer coupled proportional drive can be used only if the collector current varies in each switching period. However, the proportional drive described here in this paper can be employed even when collector current does not change continuously in a switching period.

References

1. P.R.K. Chetty, "Spacecraft Power Systems—Some New Techniques for Performance Improvement," Ph.D Thesis, Indian Institute of Science (I.I.Sc), India, 1978.
2. P.R.K. Chetty, "Dc-Dc converter maintains high efficiency," *Electronics,* January 3, 1980, pp. 159-161.

LINEAR POWER SUPPLIES

Introduction

Before the arrival of the switch mode power supplies, the linear power supplies were very famous and were used extensively. Though these are dissipative, they exhibit large bandwidth, good dynamic response and high line rejection. Even today they have been used, but only for some judiciously selected applications. Some of such applications include laboratory power supplies, isolation regulators for low level data amplifiers, logic card regulators, small instrument power supplies, airborne systems and other power supplies for digital and linear circuits.

This paper presents a power supply design working on the principle of linear regulation. The power supply described here regulates the output voltage down to zero volts.

Description of Control IC

Although there are many ICs for this application, LM723 has been considered in this practical example. This IC, as shown in Fig. 1, is a monolithic voltage regulator and it consists of a temperature compensated reference amplifier, error amplifier, series pass transistor and current limit circuitry. Additional npn or pnp pass transistors may be used when output currents exceeding 150 mA are required. Provisions are made for adjustable current limiting and remote shutdown. In addition to the above, the device features low standby current drain, low temperature drift and high ripple rejection. This IC can be used as a series, shunt, switching or floating regulator.

Practical Example

In most dc-input regulated power supplies, regulation is poor when the desired output voltage is less than the controller IC's internal reference voltage. In addition, circuit considerations usually limit the minimum reference voltage attainable and consequently the minimum output voltage possible. In this example, however, the reference voltage is brought down virtually to zero, to overcome both the problems.

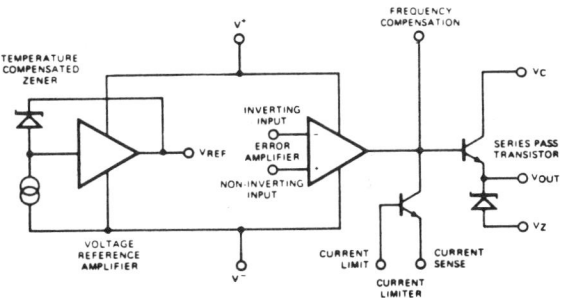

Fig. 1. Block schematic of LM723 monolithic voltage regulator.

The LM723 voltage regulator shown in Fig. 2, reproduced from [1], which provides 12 volts at 1 ampere, must be biased with a negative voltage supply at its −Vin port (pin 5) for proper operation. This voltage is provided by the switching inverter shown within the dotted lines. The scaled down output voltage is compared with the reference voltage. The error voltage is amplified and fed to

---

Portions are reprinted with permission from *ELECTRONICS,* Jan. 19, 1978. Copyright © 1978. McGraw-Hill Inc. All rights reserved.

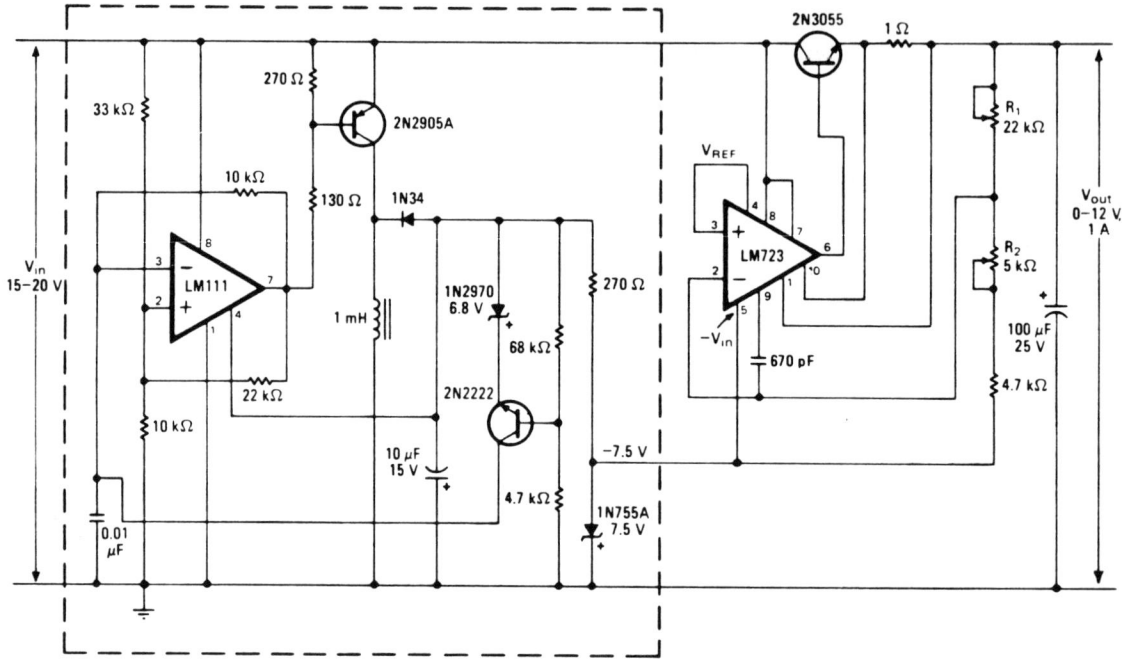

Fig. 2. Detailed dc-dc power supply schematic for regulating down to 0 volt.

NOTE: 2N3055 NEEDS HEAT SINK

the series pass transistor. The series pass transistor works as a variable resistor whose value changes as a function of the error voltage. This type of negative feedback forces the series pass transistor to drop a voltage across its collector-emitter equal to input voltage minus required output voltage and thus the output is maintained at a predetermined voltage level.

The LM111 voltage comparator is configured as an astable multivibrator that oscillates at a frequency of about 10 kilohertz. With the aid of the 1-millihenry inductor, which generates the counter-electro-motive force required to produce a negative potential from switched-input voltage, the inverter delivers a well regulated -7.5 V to the $-V_{in}$ port of LM723.

The magnitude of this bias voltage is essentially equal to that of the regulator IC's internal reference voltage, V_{ref}, appearing at pin 4, and properly biases its voltage reference amplifier. This condition in turn precipitates a condition in the amplifier whereby V_{ref} clamps to ground potential. Thus the output voltage may be adjusted throughout its maximum possible range by potentiometers R1 and R2. Although the potential of V_{ref} as measured with respect to ground has been changed, the circuit will retain the regulating properties of the LM723. Both the line and load regulation of the supply are 0.4%.

Reference

1. P.R.K. Chetty and A. Barnaba, "DC-DC Power Supply Regulates down to 0 volt," *Electronics*, USA, Jan. 5, 1976.

IMPROVEMENTS TO POWER SUPPLIES

Introduction

Simple changes like adding a component appropriately can improve the performance of a power supply. Two such improvements are presented here, one for a linear power supply and the second one for a switch mode power supply.

Practical Example-1

In power supplies, the series-pass transistors and current sensing resistors are the power dissipating elements, and despite current limiting, high-power IC regulators can experience excessive power dissipation when their outputs are shorted. This situation arises because each series-pass transistor must dissipate the power generated by the full input voltage at a current slightly greater than that for full load. Such dissipation can easily be three times the worst case value for normal, full-load operation.

To avoid this situation, voltage regulators incorporate foldback current limiting. The short circuit current depends on the current sensing resistor, and to achieve a low value of this quantity for a fixed full-load current, a larger sensing resistor is needed. But that component again dissipates more power and reduces efficiency during normal operation. In addition, it requires a heat sink.

A diode connected in series, as shown in Fig. 1, with the current-sensing resistor improves the circuit by allowing use of a smaller sense resistor and thereby reducing the power dissipation. For this example circuit, the savings are:

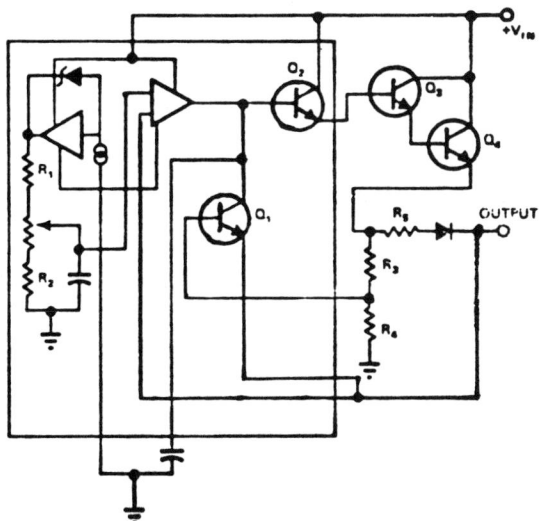

Fig. 1. Addition of a diode improves current foldback of a power supply.

Standard circuit: Rsense = 15 ohm, 3.75 W dissipation
Improved circuit: Rsense - 1 ohm, 0.60 W dissipation

Practical Example-2

As mentioned above, in power supplies, series pass transistors and current sensing resistors are the power dissipating elements. Though the introduction of switch mode power supplies, reduced the power dissipation in series pass transistors, the dissipation in current sensing resistors remained unchanged. Most of the regulators have a Vsense of about 0.7 V which is the base-emitter voltage of a transistor. For a fixed output current, the power dissipation in, size and weight of current sensing resistor depend upon (proportional to) the Vsense. A diode connected as shown in Fig. 2 effectively brings down the Vsense from 0.7 V to 0.4 V and thus reduces the dissipation, size and weight of the current sensing resistor. This in turn increases the regulator efficiency.

In view of comparing the conventional circuit with the improved circuit, the conventional circuit is considered first. In this example, the regulator has been designed for an output of 5 V and a load current of 2 A with an efficiency of 80%. As the Vsense is 0.7 volt, the Rsense shall be Vsense/Io or 0.35 ohm. The power dissipation in Rsense is given by $(V_{sense})(I_o)$ or 1.4 watts. As this power supply has an efficiency of 80%, the power dissipated in the power supply is 2.5 watts (input power minus output power). This dissipation is due to current sensing resistor and due to other components like series pass transistor, IC, passive components, etc. The power dissipation due to other than current sensing resistor is equal to 2.5 W − 1.4 W = 1.1 W.

Now the improved power supply is considered. From Fig. 2(C), V_{be} is equal to $(I_o)(R_{sense})$ plus the diode drop (Vd), or the V_{sense} is equal to $(V_{be} - V_d)$.

Portions are reprinted with permission from *Electronic Engineering*, UK, April 1980.
Portions are reprinted with permission from *Electronic Design News (END)*, October 5, 1978.

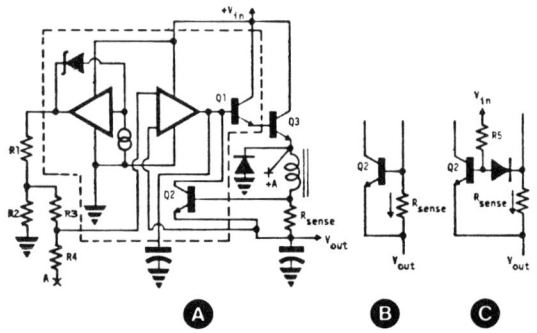

Fig. 2. Addition of a diode improves the efficiency of a regulator.

By choosing a Germanium diode, whose forward voltage drop is about 0.3 V, Vsense will be equal to 0.7 V − 0.3 V = 0.4 V. The power dissipation due to current sensing resistor in the improved circuit is 0.8 W. The power dissipation due to additional diode and resistor is 0.1 W and the power dissipation in remaining power supply is 1.1 W. Thus the total power dissipation in the improved circuit is 2.0 W. Therefore the efficiency of the improved circuit, given by (output power/(output power + dissipated power)), is 83.3%.

Table 1. Comparison of the Conventional and the Improved Circuits.

Parameter	Conventional Circuit	Improved Circuit
Value of Rsense	0.35 ohms	0.2 ohms
Dissipation in Rsense	1.4 W	0.8 W
Size (relative)	1.0	0.6
Weight (relative)	1.0	0.6
Efficiency	80.0%	83.3%

The comparison characteristics of the conventional and improved circuit are shown in Table 1.

Thus the size, weight and the power dissipation of R_{sense} of conventional circuit has been bought down and the efficiency has been increased just by the addition of a diode and a resistor.

References

1. P.R.K. Chetty, "Add a diode to improve current foldback," *EDN* magazine, October 5, 1978.
2. P.R.K. Chetty, "Add a diode to improve the efficiency of a regulator," *Electronic Engineering,* April 1980.

Chapter 6

ICs for Switch-Mode Power Supplies

Control ICs for Switch Mode Power Supplies 132

IC Timers as Controllers for Switch-Mode Power Supplies 140

CONTROL ICs FOR SWITCH MODE POWER SUPPLIES

1.0 Introduction

The benefits obtained from switch-mode power supplies (SMPS) have become universally recognized by power systems engineers in the past several years. However, gains in efficiency and reduction in weight have been accompanied by the complexity of the discrete circuitry required to provide the proper signals for adequate control of the switching transistors, an escalating component count, and a decrease in reliability and predictability of performance. Thus, the development of SMPS had been slow. In an effort to solve these problems, many component manufacturers have introduced new devices designed specifically for switching power supply applications. These include faster power transistors with improved safe operating area (forward and reverse), fast recovery switching power diodes, low ESR capacitors, low-loss high-frequency cores and monolithic IC control devices. The control devices offer the advantages of compactness, accuracy, reproducibility, higher performance through reduction of parasitics, and the economies of mass production.

Thus, in this paper the control integrated circuits (ICs) for free-running as well as driven-type SMPS have been described. Comparison of various ICs in each category has been made. Then special power supply control ICs have been described. Also included are the protection and instrumentation ICs for SMPS.

2.0 Control ICs

Switching power supply implementation becomes easier with the help of ICs. Basic building blocks of SMPS are a precision reference source, an error amplifier, a differential voltage comparator, a driver stage and a power stage. Each of these circuits, except for the power stage, have been available in integrated circuit form for several years. Although each IC offers the benefits of reduced physical size, greater reliability, and increased performance, the complexity increases due to the large number of parts and their differential temperature, etc. Hence, it is advantageous to large-scale integrate all these parts into a single LSI linear IC.

Following the footsteps of the evolution of SMPS, namely, first free-running (ripple) switching regulators have been developed and became famous. Then the driven-type switching regulators followed. First the ICs for free-running regulators were made, then for the driven types. In the case of ripple regulators, the switching frequency varies over a range as a function of input voltage and output load. Hence, power supply filter design cannot be optimized. Still the ripple regulators are preferred because of their reduced complexity, size and cost compared to driven-type SMPS for some applications. Thus both types of ICs are important for power supplies. Hence, the following two sections deal about these two types of ICs in detail.

2.1 Control ICs for Driven Type Switch-Mode Power Supplies

An ideal control IC for driven type SMPS should include not only the elements necessary for normal pulse-width modulation operation, but also contain as many features as possible as given below:

Supply Operation to 40 volts
Highly Stable Temperature Compensated Reference Source
Sawtooth Oscillator with Deadband Control
Synchronizability with external Clock
Programmable Softstart
Error Amplifier with wide Common Mode Range
PWM Comparator with hysteresis
Pulse Steering Flip-Flop
Dual Source/Sink Output Drivers with Short-Circuit Protection
Double Pulse Suppression Logic
Symmetry Correction Capability
Current Limit with wide Common Code Range
Cycle by Cycle Current Limiting & Shutdown Circuitry
Feed Forward Control
Under-Voltage Lockout
Over-Voltage Protection

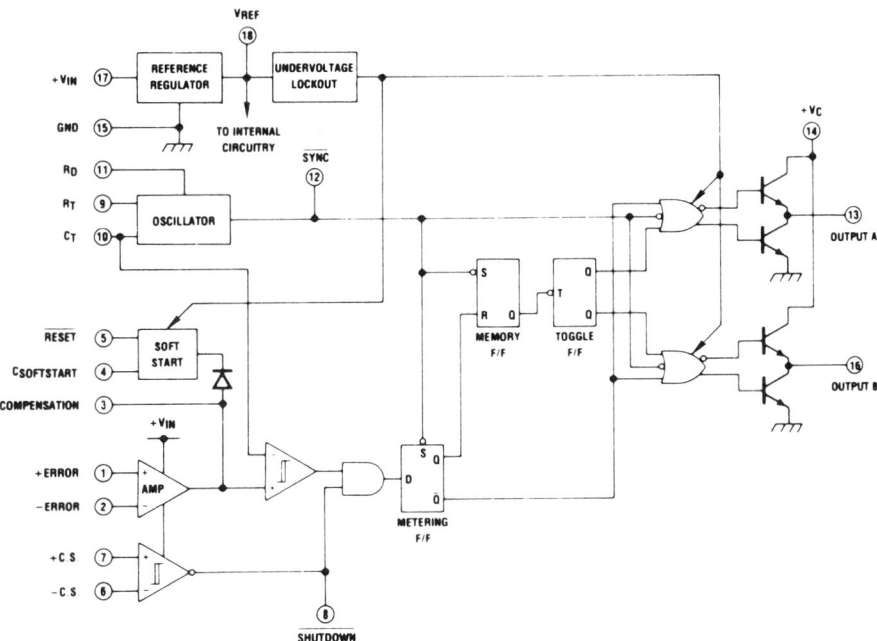

Fig. 1. Schematic of a control IC (SG1526) for driven type switch-mode power supplies (courtesy of Silicon General).

TTL/CMOS Compatible Logic
Remote ON/OFF

Typical switching frequency of operation is 50 kHz to 500 kHz and these ICs shall have a drive capability in the range of 100 mA to 200 mA. Nowadays there are many ICs which have most of the above features and one such control IC is shown in Fig. 1.

Comparison of above mentioned control IC features of some of the ICs is carried out and is presented in Table-1. A PLL circuit for synchronization is one of the special advantages offered by some ICs. Depending upon the application, a trade-off has to be carried out giving proper weightage to different factors/features and then a suitable IC has to be selected.

2.2 Control ICs for Free-Running Switch-Mode Power Supplies

As the growth of these type of ICs has been short lived, no emphasis has been made to optimize or improve these ICs. No real attempt has been made to add additional features. These ICs do not have any clock or synchronizing provisions. Following are some of the common features of these ICs, and Fig. 2 shows the schematic of one such control IC.

Supply Operation to 40 volts
Highly Stable Reference Source
Error Amplifier Comparator with hysteresis
Current Limit
Inhibit/Shutdown Circuitry
Under-Voltage Lockout
Fixed ON Time Control
Fixed OFF Time Control
Flyback Diode Inclusion

Comparison of above mentioned features of some of the ICs is carried out and is presented in Table-2. Depending upon the application, a trade-off has to be carried out giving proper weightage to different factors/features and a suitable IC has to be selected.

2.3 Special Purpose Control ICs

In addition to main ICs for control, there are

Table 1. Control ICs for Driven Type Switch-Mode Power Supplies. (continued thru page 137)

IC Designation	UA494*	TL593$	TL595@	UC1840**	ULN8160 $*
IC name	PWM Control Circuit	PWM Controller	PWM Controller	PWM Controller	SMPS Controller Circuit
Made by	Fairchild	TI	TI	Unitrode	Sprague
No. of pins	16	16	18	18	16
Freq. (kHz)	1-300	—	—	—	100
Vcc (volts)	42	—	—	—	18
Ic, Peak Collector Current (mA)	250	200	200	20	40
Multipulse Suppression	No	No	—	—	Yes
Sync Input#	NP, but possible	NP, but possible	NP, but possible	No	Yes
Shutdown	NP*#	NP	—	Yes	Yes
Reset	NP	NP	—	Yes	—
Soft-start	NP*#	NP	—	Yes	Possible
Common Mode Range (V)	−0.3 to −2	−0.2 to −2	—	1-75	—
FT (kHz)	650	—	—	2000	3000
Outputs	Double	Double	—	Single	Single
Under Voltage Lockout	—	Yes	Yes	Yes	Yes
Output Control	Yes	Yes	—	NA	NA
Reference (Volts)	—	5+/−1%	5+/−1%	5	3.75
Current Sense Op Amp/Extra Error Amp	Yes	Yes	Yes	Yes	Yes

Table 1. Control ICs for Driven Type Switch-Mode Power Supplies. (continued thru page 137)

IC Designation	LAS3840$$	LAS3800@@	NE5561$@	ZN1060@*	ZN1066$#
IC name	Monolithic Switching Regulator	Monolithic Switching Regulator	SMPS Controller	Monolithic Switching Regulator Control Ckt	Switching Regulator Control & Drive Unit
Made by	Lambda	Lambda	Signetics	Ferranti	Ferranti
No. of pins	16	16	8	16	24
Freq (kHz)	1-500	1-150	100	50-100	500
V_{CC} (volts)	12-40	12-40	18	5	5
I_C, Peak Collector Current (mA)	200	500	40	40	60
Multipulse Suppression	Double Pulse	Double Pulse	Yes	Yes	Yes
Sync Input#	Yes	NP, but Possible	No	Yes	Yes
Shutdown	Yes	—	No	Yes	Yes
Reset	—	—	No	—	—
Soft-start	Yes	—	NP, but Possible	Yes	NP, but Possible
Common Mode Range (V)	0-11	—	—	—	1-4
FT (kHz)	—	—	3000	—	—
Outputs	Double	Double	Single	Single	—
Under-Voltage Lockout	Yes	No	No	Yes	—
Output Control	No	No	NA	NA	—
Reference (V)	1.65	—	3.75	3.73	2.6
Current Sense Op Amp/Extra Error Amp	Yes	—	Yes	Yes	Yes

Table 1. Control ICs for Driven Type Switch-Mode Power Supplies. (continued thru page 137)

IC Designation	SG1524	SG1525/ SG1527	SG1525A/ SG1527A	SG1526
IC name	Regulating Pulse Width Modulator	Regulating Pulse Width Modulator	Regulating Pulse Width Modulator	Regulating Pulse Width Modulator
Made by	Silicon General	Silicon General	Silicon General	Silicon General
No. of pins	16	16	16	18
Freq (kHz)	300	0.1-400	0.1-500	0.001-400
V_{CC} (volts)	8-40	8-40	8-40	8-40
I_c, Peak Collector Current (mA)	100	100	100	100
Multipulse Suppression	No	No	Yes	Yes
Sync Input#	No, but possible	Yes	Yes	Yes
Shutdown	Yes	Yes	Yes	Yes
Reset	No	No	No	Yes
Soft-start	No, but possible	Yes	Yes	NP
Common Mode Range (V)	1.8-3.4	1.5-5.2	1.5-5.2	0.0-5.2
F_T (kHz)	3000	2000	2000	1000
Outputs	Pushpull	Pushpull (Totempole)	Pushpull (Totempole)	Pushpull (Totempole)
Under-Voltage Lockout	No	No	Yes	Yes
Output Control	NA	NA	NA	NA
Reference (V)	5.0	5.1	5.1	5.0
Current Sense Op Amp/Extra Error Amp	Yes	No	No	Yes

NOTES For Table 1

*   TL495 provides identical functions found in TL494. In addition the TL495 contains an on-chip 39V zener diode for high voltage applications where Vcc is greater than 40V and an output steering control that overrides the internal control of the pulse steering flip-flop. TL495 is 18 pin IC. Pin 13 of TL494 is sensitive to noise pick up; as Vcc drops below 8V, the two outputs can be ON simultaneously which may be destructive in push-pull applications.

$ TL594 provides identical functions found in TL593, except for current limit amplifier. TL594 contains two error amplifiers for design flexibility. TL594 is designed to replace TL494.

@   Has internal 39V zener and also it has an output steering control.

\#   The power supply shall be running at its own oscillator frequency in the absence of sync input and shall run/operate at sync frequency when the sync input is present.

**   Has feed forward control input, over voltage protection, maximum duty cycle limiting requires three extra components, clock effectively uses four components, approximately totem-pole output.

$$ LAS3840 has thermal shutdown, double pulse suppression, sawtooth oscillator with overcurrent frequency shift, dynamic volt-time symmetry correction in double ended systems. LAS3820 has input voltage limitation of 20V compared to 40V limit for CAS3840.

@@   Has frequency shift control.

*#   Pin 4 can be used for soft-start as well as for shut down.

*$ Feed forward control, also has demagnetization, high voltage protection.

$@   The device in intended for low cost SMPS applications where extensive housekeeping functions are not required.

$#   Pulse steering flipflop.

@*   Has over voltage protection, demagnetization antisaturation protection.

FT   Unity gain bandwidth.

NA   Not applicable.

NP   No provision.

some special-purpose control ICs like the microprocessor power-supply control IC, micropower power supply control IC, etc., which are described here.

2.3.1 Microprocessor Switch-Mode Power Supply Control ICs

TCA5600/TCF5600 is a versatile power-supply control circuit for microprocessor based systems and mainly intended for automotive applications and battery-powered instruments. To cover a wide range of applications, the device offers high circuit flexibility with a minimum of external components.

This IC includes a temperature-compensated voltage reference source, on chip dc-dc converter, programmable and remote-controlled voltage regulator, fixed 5.0 V supply voltage regulator with external pnp power device, undervoltage detection circuit, power-on reset delay and watch-dog feature for orderly microprocessor operations.

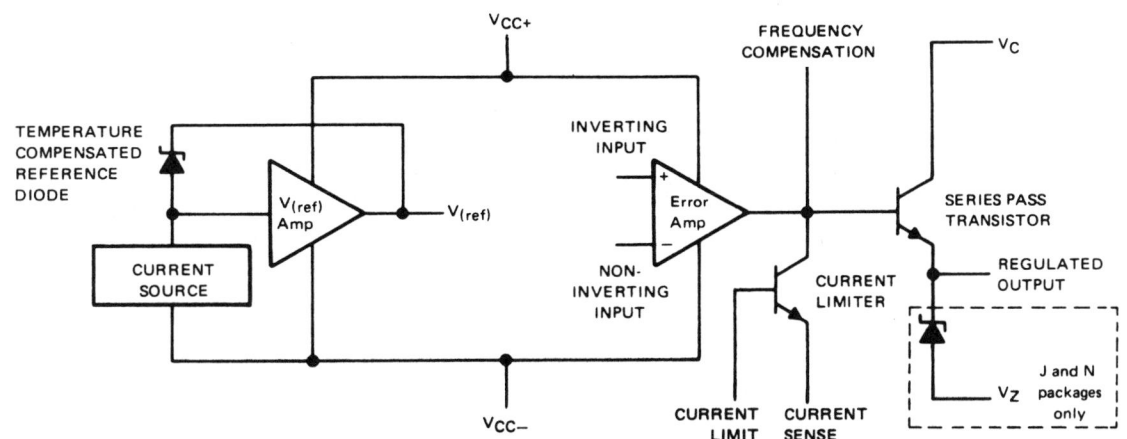

Fig. 2. Schematic of a control IC (UA723) for free-running switch-mode power supplies (courtesy of Texas Instruments Inc.).

Table 2. Control ICs for Free-Running Switch-Mode Power Supplies.

IC Designation	UA723	SG1532	TL497A	UA78S40	LM105	LM104
IC Name	Precision Voltage Regulator	Precision General Purpose Regulator	Switching Voltage Regulator	Universal Switching Regulator Subsystem	Positive Voltage Regulator	Negative Voltage Regulator
Made by	TI	Silicon General	TI	Fairchild	TI	TI
No. of pins	10 DIP / 10 CIR / 14 DIP	10 CIR / 14 DIP	14 DIP	16 DIP	8 CIR / 8 DIP	10 CIR / 14 DIP
Vcc (volts)	9.5-40	4.5-50	4.5-12	2.5-40	8.5-50	8-50
Reference (V)	7.15	2.5 (typ)	1.2	1.3	1.7 (typ)	—
Switch current/output current (mA)	150	250	500	1500	12-45	20
Common mode input	—	—	*	−0.3V – V+	*	*
Output (V)	2-37	2-38	Vin + 2 to 30	1.3-40	4.5-40	0.015-40
Fixed ON time	No	No	Yes	—	No	No
Current Limit Provision	Yes	Yes	Yes	—	Yes	Yes
Inhibit/Shutdown	Possible	Possible	Yes	—	Yes	Yes
Flyback Diode Included	No	No	Yes	No	No	No

*Reference is tied to one of the computer inputs internally.

2.3.2 Micropower Switch-Mode Power Supply Control ICs

4191, 4192, and 4193 are the industry's first monolithic micropower switching regulators available in an 8-lead mini-DIP and designed specifically for battery-operated instruments. They each contain a 1.3 V temperature-compensated bandgap reference, adjustable free-running oscillator, voltage comparator, low-battery detection circuitry, and a 150 mA switch transistor with all of the functions required to make a complete low-power switching regulator.

3.0 Driver ICs

As power systems engineers gained experience in applying the control ICs, the gap between the output power capabilities of the control ICs and the drive levels required by the semiconductor switches became apparent. Most power-supply configurations where specialized driver functions could be successfully implemented with monolithic technology are a dual-output driver and a high-current floating switch driver.

Figure 3 shows the schematic of the SG1627 dual-output driver IC. This employs totem-pole driver configuration with externally programmable current sourcing and has 500 mA source capability. Both inverting and non-inverting logic inputs are available, and may be driven by either an open collector control circuit or (with a diode) by TTL logic. Connections to the high-current output transistors are brought out separately, allowing maximum flexibility when interfacing with standard bipolar transistors, the new VMOS power FETs, and transformers.

SG1629 is a high-current floating switch driver

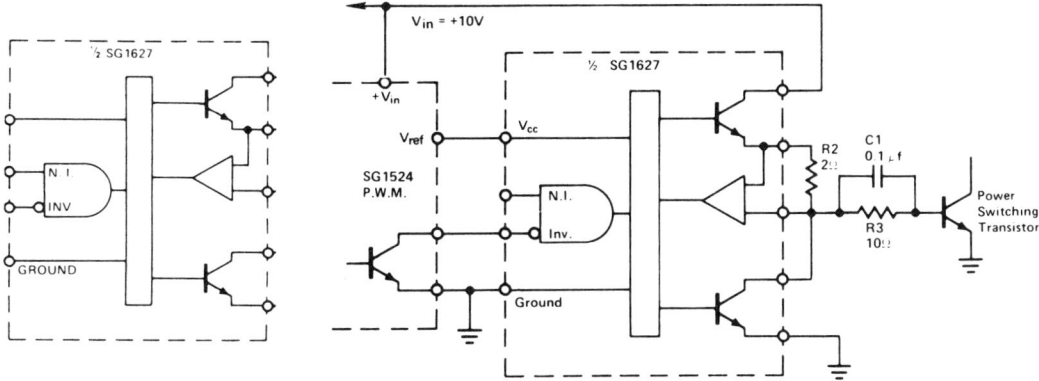

Fig. 3. Schematic of a dual output driver (SG1627) for switch-mode power supply application (courtesy of Silicon General).

that has been designed to provide an interface between a drive transformer secondary winding and a high-power switching transistor with adequate turn-on and turn-off drive capability. It does not require any external power supply as it develops all the power for both turn-on and turn-off from the drive transformer and an external storage capacitor. This circuit also contains the capability for constant-current drive operation with a similar type of current sensing circuit and an external current sensing resistor.

4.0 ICs for Protection, Monitoring, Etc.

In addition to the ICs for obtaining the regulation, most power systems require additional circuitry for monitoring satisfactory performance, providing protection in the event of a fault condition, etc. These requirements led to the development of output supervisory control circuits. Beginning with a simple over-voltage sensing circuit, these devices range through more versatile and accurate single function units, to all inclusive devices that contain sensing circuits for both over and under voltage conditions, current sensing and SCR crow-bar firing, logic outputs, and an accurate independent reference generator. Figure 4 shows the schematic of one such power-supply output supervisory control circuit IC and following is the list of some of such ICs.

SG3523	Over-voltage Sensing Circuit
SG1542	Voltage Sensing and Protection Circuit
SG1543	Power Supply Output Supervisory Circuit
SG1544	Low Voltage Supervisory Circuit
SG1547	Quad Power Fault Monitor
SG1549	Current Sense Latch
MC3523/MC3423	Overvoltage 'Crow-Bar' Sensing Circuit
MC3425/MC3525	Overunder Voltage Protection Circuit
MC34061/MC34062	Overvoltage Protection Circuit
MPC2005	Overvoltage and Temperature Protector Circuit

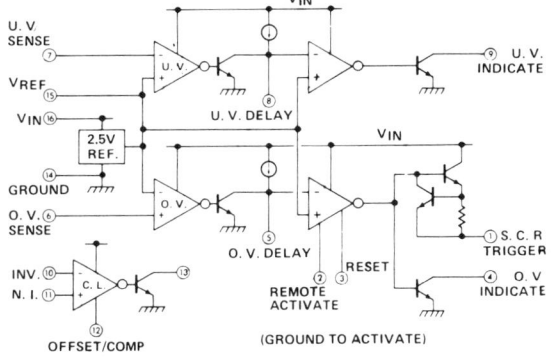

Fig. 4. Schematic of a supervisory control IC (SG1543) (courtesy of Silicon General).

References

1. Silicon General linear Integrated Circuits—Product Catalog 1980.
2. Siemens, "Switched-Mode Power Supplies Using the TDA4600," Application Note, Components Group.
3. Siemens, "Control ICs TDA4700/4718 for Switched-Mode Power Supplies-Function and Application Note," Application Note, Components Group.
4. Preliminary Data Sheets, Fairchild—A Schlumberger Company.
5. *The Linear Control Circuits Data Book for Design Engineers,* Texas Instruments Incorporated.
6. Many other Data Catalogs.

IC TIMERS AS CONTROLLERS FOR SWITCH MODE POWER SUPPLIES

Introduction

Not only the regulator ICs or pulse-width modulating regulator ICs can be used for switch-mode power supplies, but other ICs as well. A regulating IC is not a must for designing a switch-mode power supply, which is proved in this paper by advantageously employing a 555 Timer IC. Thus, this paper presents three practical examples wherein the timer is used as a controller for switch-mode power supplies. The Timer IC is manufactured by many different manufacturers. Most of them call it as 555 Timer, but Motorola calls it as 1455 Timer although it is similar to the 555 Timer.

Description of Timer IC

As the 555 Timer IC is used as the controller for switch-mode power supplies, it is briefly described here. The timer as shown in Fig. 1, consists of two voltage comparators, a bistable flip-flop, a discharge transistor, and a resistor divider network. The resistive divider network sets the comparator levels and since all three resistors are of equal value, the threshold comparator (A1) is referenced internally at 2/3 of supply voltage (VCC) level and the trigger comparator (A2) is referenced at 1/3 of VCC. The outputs of the comparators are tied to the bistable flip-flop (A3). When the trigger voltage moves below 1/3 of VCC, the comparator changes state and sets the flip-flop driving the output to a high state. When the threshold comparator voltage exceeds 2/3 of VCC, the threshold comparator resets the flip-flop, which in turn drives the output to a low state and turns-on the discharge transistor.

In switching regulator applications, the Timer IC is configured as an astable multivibrator. Initially, when the timer is powered, the capacitor C is allowed to charge and the timer output is HIGH. The threshold comparator monitors and compares the capacitor voltage as it charges and when it reaches 2/3 of VCC, it changes the timer output to go to LOW and turns-on the discharge transistor. Now the trigger comparator monitors and compares the capacitor C voltage as it discharges and when it decreases to 1/3 of VCC, it changes the timer output to go to HIGH and turns-off the discharge transistor. Now the capacitor C is allowed to charge ... the cycle repeats. From Fig. 1, it is clear that charging is controlled by $R1 + R2$, while discharge is controlled only by R2. Thus the approximate fre-

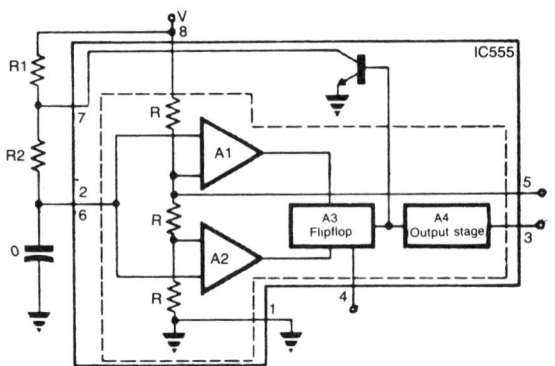

Fig. 1. Block schematic of a timer IC.

Portions are reprinted with permission from *Electronic Design News* (EDN), Jan. 5, 1976.

Portions are reprinted with permission from ELECTRONICS, Nov. 13, 1975. Copyright © 1975. McGraw-Hill Inc. All rights reserved.

quency of operation is given by $fs = 1.44/(R1 + 2.R2)(C)$.

Timer IC as Switching Regulator Controller

As mentioned above, the 1/3 of VCC and 2/3 of VCC are generated internally by the Timer IC by employing three equal resistors in series. As the 2/3 of VCC reference point is brought out as pin 5 (control voltage), the charge or discharge periods can also be controlled by overriding the internal 2/3 of VCC by applying appropriate voltage at pin 5 in a predetermined timely fashion. Thereby the frequency and effective duty ratio can be controlled. Also pin 4, which is external reset, can be used to control the timer output to stay LOW as long as this pin voltage is LOW, overriding the timer internal control. Thus by using either pin 4 or pin 5, the duty ratio of the timer can be controlled. The output stage of the timer is of totem-pole design and has sink or source capability of 200 mA. This signal can be used to drive the power transistor of switching dc-dc converter. In closed-loop regulator mode, the regulator output is compared with a reference voltage and the amplified error voltage is used as control voltage at pin 5. Thus, the duty ratio of the timer output is varied to maintain the output voltage of the regulator at a predetermined level. Reset pin can be used to add a protection feature.

Practical Example 1

Figure 2 shows a current step-up converter regulator, which is reproduced from.[1] When the timer output is HIGH, transistor Q2 is turned-on and therefore pass transistor Q3 is turned-on. Collector current from Q3 flows through inductor L into the load and the filter capacitor. When the output of the timer goes LOW, the transistors turn-off. Diode D commutes the current flowing through the inductor when Q3 switches-off. If there was no feedback circuit, the output voltage would depend upon the input voltage and the duty cycle.

The feedback circuit consists of R4, zener diode DZ2, transistor Q1, and R3. Whenever the output voltage exceeds (Vz2 + Vbe1), Q1 turns-on and drives the reset pin of the timer LOW. The transistors Q2 and Q3 therefore stay off, allowing the output voltage to decrease. Thus the output voltage, V_{out}, is maintained at a voltage approximately equal to $(V_{z1} + V_{be1})$.

The performance of the current step-up converter regulator is as follows:

Input Voltage, V_{in} = 15 V
Output Voltage, V_{out} = 8.4 V

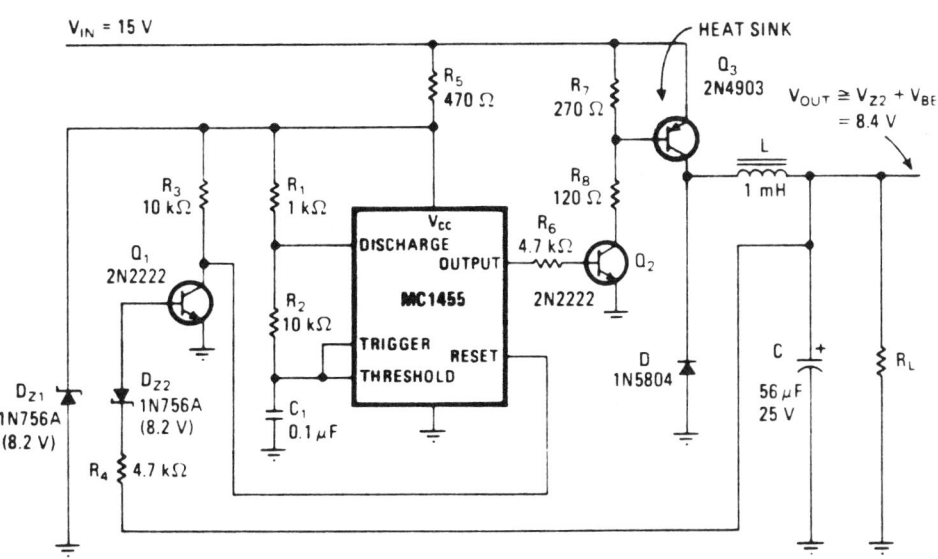

Fig. 2. Current step-up converter regulator using timer IC as controller.

Load Current, I_{out}	= 300 mA
Ripple, I_r (for I_{out} = 300 mA)	= 5 mA
Load Regulation (for V_{in} = 15 V and I_{out} = 0-300mA)	$</= 0.5\%$
Line Regulation (for V_{in} = 15-25 V and I_{out} = 300 mA)	$</= 2.5\%$

Modern Timer ICs operate at a frequency of 500 kHz or higher. The maximum operating voltage (VCC, at pin 8) of the timer is 16 volts, but here its VCC is clamped at 8.2 V by a zener diode Dz1. The input voltage therefore can have any value within the ratings of the pass transistor and the filter capacitor.

Practical Example 2

Figure 3 shows a polarity reversing regulator or buck-boost regulator reproduced from.[1] This circuit differs from Fig. 2 in the arrangement of L, C, D, and the feedback components. When Q3 switches-off, the commutating current in L charges C to produce an output voltage that is negative with respect to ground. This voltage is applied to the anode of Dz2 through limiting resistor R4. Whenever the output voltage is more negative than $-(V_{z2} + V_{be1})$, the timer reset goes LOW, allowing the voltage across the capacitor to become less negative. Because of this closed-loop action, the output voltage of this circuit is maintained at approximately $-(V_{z2} + V_{be1})$. This circuit can provide an output voltage equal to, less than, or greater than the input voltage.

The performance of the circuit in Fig. 3 is as follows:

Input voltage, V_{in}	= +15 V
Output voltage, V_{out}	= -19.4 V
Load Current, I_{out}	= 300 mA

Ripple and regulation are the same as in practical Example 1.

Practical Example 3

Figure 4 shows a switching regulator reproduced from[2] and is similar to Fig. 2, but for an additional protective feature. Referring to Fig. 4, capacitor C1 charges to 2/3 of V_{in} through R1 and R2, and discharges to 1/3 of V_{in} through R2 when there is no external voltage at pin 5. Thus the timer will retrigger itself, producing a square-wave output. This square wave, amplified by Q1, is fed to transistor Q2. As long as the timer output is HIGH, Q2 will be on and driving current into R8

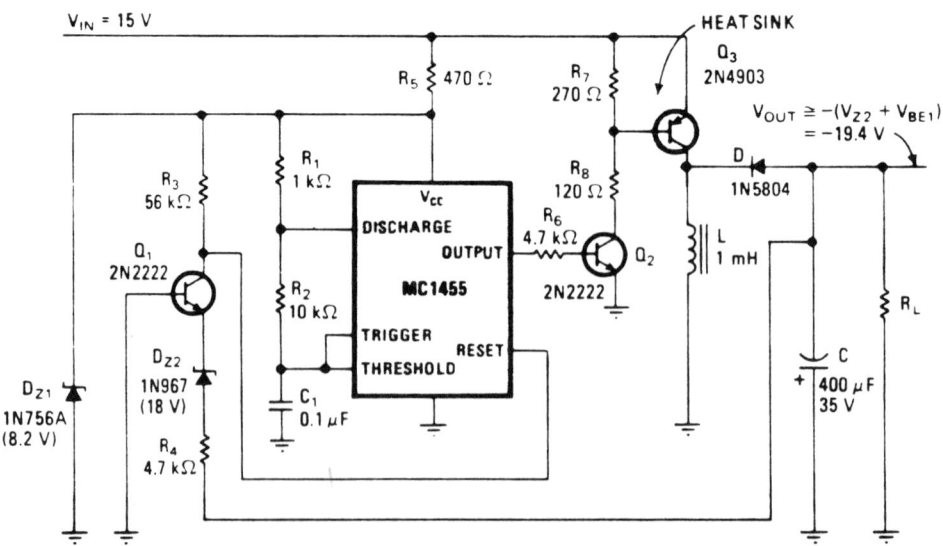

Fig. 3. Polarity reversing or buckboost regulator using timer IC as controller.

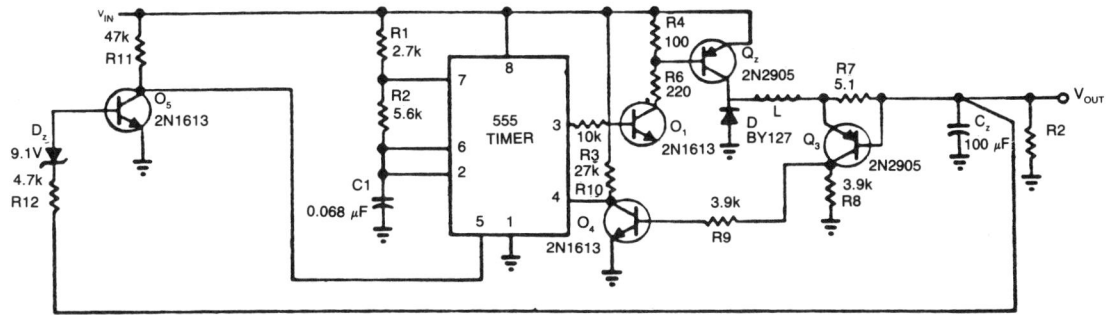

Fig. 4. Switching regulator with current foldback using timer IC as controller.

(load) and C2 through inductor L. When Q2 turns-off, the diode D commutes the current flow through the inductor and the energy stored in L and C2 is available to supply the load. The output voltage is fed to a simple comparator formed by Q5, Dz, R11 and R12. Q5 conducts when the output voltage exceeds the zener voltage plus the base-emitter voltage. Since the collector voltage of Q5 is fed to modulating input (pin 5), the pulse width of the generated square wave is modulated to maintain the output voltage at a predetermined level, i.e., at about $V_{z2} + V_{be1}$. An approximate relation between V_{in} and V_{out} can be described as:

$$V_{out} = (t_{on} \cdot V_{in}) / (t_{on} + t_{off})$$

R7 is the current-sensing resistor. When the load current increases to a level such that the voltage drop across R7 turns Q3 on, Q4 will be driven into saturation. As the collector of Q4 is tied to pin 4, it resets the timer, bringing its output to LOW, thereby Q2 turns-off. Thus, with the timer reset, no voltage develops across R8 and Q4 is turned-off, enabling the timer, and changing its output to go HIGH. If an overload condition still exists, Q3 and Q4 will again be turned-on and reset the timer.

This closed loop chain reaction continues as long as an overload condition exists. If the overload condition increases, the voltage and current will both decrease initiating the foldback action.

With a 15 V input, the circuit delivers a 10 V, 100 mA output with line and load regulation of 0.5% and 1%, respectively. Foldback action will commence at a current value equal to Q3's $V_{be(sat)}$ divided by R7.

Conclusions

It has been shown that for designing switching regulators other than regulator ICs can be used. Use of a Timer IC for switch-mode power supplies has been demonstrated with three practical examples.

References

1. P.R.K. Chetty, "IC Timers Control DC-DC Converters," *Electronics,* Nov. 13, 1975.
2. P.R.K. Chetty, "Put a 555 Timer in your Next Switching Regulator Design," *EDN Magazine,* USA, Jan. 19, 1978.

Chapter 7

Spacecraft Power Systems

Spacecraft Power Systems   146

Improved Power Conditioning Unit for Regulated Bus Spacecraft Power System   151

SPACECRAFT POWER SYSTEMS

1.0 Introduction

The continuous source of energy in space is solar radiation and is free and abundant. Hence, solar cells are employed to convert solar radiation into electrical energy. However, when a spacecraft is in shadow or eclipse, solar cells do not produce any power. Hence, to have continuous source of power, some electrical energy is stored during sunlit period which is used during eclipse. Thus, the power requirements of most of the spacecraft, are met by the combination of solar cell and hermetically sealed alkaline batteries, which proved its promise in its performance and high reliability. While on the sunlit side of the earth, the solar cells power the loads as well as recharge the battery, which will take over and power the loads during the orbital eclipse.

The important building blocks of a typical spacecraft power system are shown in Fig. 1. These building blocks can be interconnected in different ways to result in different spacecraft power systems. In any type of spacecraft power system, the outputs of solar cell array and the storage battery are to be conditioned so as to match with requirements of the various subsystems. The battery has to be charged from the solar cell array during the orbital day and discharged to provide power during the orbital night or when the load demand exceeds the solar cell array capability. All these functions are carried out by means of the power conditioning and control systems (PCCS).

2.0 Energy Sources

As mentioned above, solar radiation is the main energy source in space and solar cells are used to convert it into electrical energy. Storage batteries are used during eclipse. Thus, the following section includes description of solar cells, storage cells, and batteries.

2.1 Solar Cells

Solar cell is essentially a large-area shallow-junction device with relatively low open circuit voltage and short circuit current, and with base resistivity of, typically, 10 ohm-cm. The cells shall be capable of withstanding repeated thermal cycling in vacuum between +100 degrees C and −200 degrees C. The solar cells convert solar energy into electricity at about 14.5% efficiency at 25 degrees C under Air Mass Zero (AMO) condition and this decreases at the rate of 0.6% degree C with rise in temperature. Further, the cells show degradation in radiation and micro-meteorite environment, which is minimized using suitable protective glass attached to the cells with special ultra-violet radiation resistant adhesives or using solar cells with integrated quartz or sapphire cover glasses. The cover glass incorporates an antireflection coating on the front surface, a multilayer UV rejection filter at the back surface and has a transmittance of better than about 94%. Temperature of the cells has to be kept low to achieve better performance and the cover glasses act as a filter cutting down the total energy absorbed by the cell and achieve this to a good extent. Wrap around contact cells are preferred in view of the simplicity of interconnection. Gold plated kovar buss is used to connect the top and bottom ohmic contacts of the cells.

2.1.1 Solar Cell Array

The solar cells are combined into small modules

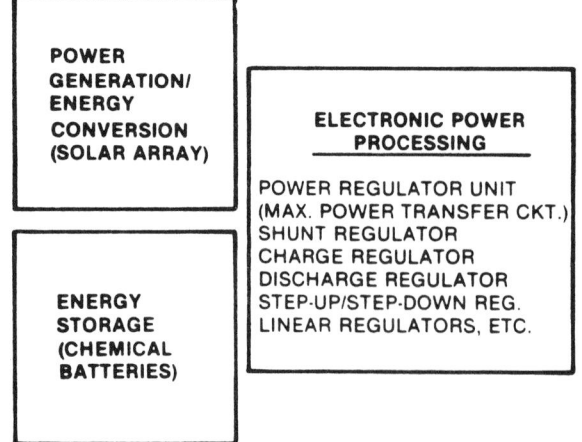

Fig. 1. Building blocks of a typical spacecraft power system.

and then into big arrays to provide power. The required voltage level is achieved by connecting the current matched cells in series and the required current level is achieved by connecting voltage-matched cell-strings in parallel. Isolation diodes are used properly in these interconnections to avoid the effects of failed cells as well as shadow effects of the projecting objects.

Flat cell modules are preferred to shingled cell modules because of the ease of replacing broken cells, more freedom in series and parallel interconnection, better thermal properties and stronger bonding to the array. Though there is a penalty paid in the reduction of active area owing to the bus bar of each cell being exposed, it is compensated by a high packing factor.

2.2 Storage Cells and Batteries

In any spacecraft power system that uses solar radiation, the storage battery is the main source of continuous power, as it responds to peak and eclipse demands of power. The required voltage level is achieved by connecting the current-matched storage cells in series and the required current level or ampere-hour rating is achieved by properly selecting the current rating of the cells connected in series and using more than one battery in parallel.

2.2.1 Low Earth Orbit Vs Geosynchronous Orbit

The number of eclipses as well as the eclipse duration depends upon the orbit altitude and inclination. In the case of the spacecraft at low altitudes, where a large proportion of time is spent in shadow, the solar cell array must be capable of generating nearly twice the average design load. On the other hand, a spacecraft at synchronous altitude is in shadow for only 5% of the time and correspondingly about 10% over design of the array is all that is required. Also the temperature excursions are large in synchronous altitude (compared to low earth orbit), although the temperature cycles are less. The eclipse seasons in a geosynchronous orbit occur twice per year, viz., in spring and autumn. Each eclipse lasts for 45 days with a maximum shadow time of 1.2 hours/day. Thus, charge and discharge cycles for any storage battery on board a spacecraft in the above orbit will be about 90 per year, as the battery is charged during sunlit portion of the orbit and discharged during the eclipse. But, in the case of low earth orbit spacecraft, the number of eclipses increases as the altitude decreases. Typically, for a 600 km orbit, there will 15 eclipses per day with a maximum shadow time of 36 minutes, for every orbit period of 96 minutes. Thus, charge-discharge cycles in this case will be about 5500 per year. Several times in a year the spacecraft will be in continuous sunlight for particular orbits (like polar orbits with a particular inclination) for long periods during which the daily average solar cell array power exceeds the average power demand. This extra power can be optimally utilized only if the battery is capable of

Table 1. Characteristics of Storage Cells.

Cell type	WH/KG	WH/Cu.M	Cycle (relative)	Space qualified
Ag-Zn	90-120	260	1	Yes
Ag-Cd	40-60	120	26	Yes
Ni-Cd	30-35	100	60	Yes
Ni-H2	45-55	40	>60	Being*
Ag-H2	70-80	50	—	To be**

* Has been flown on satellites in 12 hour and geosynchronous orbits. Ground accelerated tests show that the cycle life is better than Ni-Cd cells.
** Has been tested on ground successfully and passed the test of simulated 10 years charge-discharge cycle life for geosynchronous orbit application.

being charged at high rates. In addition, storage cells shall possess long charge-discharge cycle life, high recharge efficiency, good hermetic seals to prevent loss of electrolyte and corrosion, low weight, cost, volume, and high reliability.

2.2.2 Storage Cells-Types

Table-1 shows the important characteristics of storage cells, namely, energy density, the cycle life, whether space qualified or not, etc. It is evident from this table that the Ni-Cd cell has the optimal properties as Ni-H2 and Ag-H2 cells are yet to reach the required levels of reliability. Recent communication satellites are utilizing Ni-H2 batteries because the number of charge-discharge cycles required in geo-synchronous orbit is very low. Also, these cells are being qualified for low earth orbit satellites. The NTS-2 satellite whose orbital period is 12 hours carried Ni-H2 storage cells for the first time and since 1976 they are working successfully. Thus in the near future, Ni-H2 cells might replace the Ni-Cd cells.

Also, it appears from the present trends that the Ag-H2 cells are major competitors for future missions. The first results on the investigations of the suitability of the Ag-H2 cells seem to show that good performance can be anticipated from these, since a watt-hour density of 70-80 WH/kg can be reached. The design of Ag-H2 cells is similar to the existing Ni-H2 types.

3.0 Power Conditioning and Control Systems

In any type of spacecraft power system, the outputs of the solar cell array and the storage battery are to be conditioned so as to match with requirements of the various subsystems. The battery has to be charged from the solar cell array during the orbital day and discharged to provide power during the orbital night, or when the load demand exceeds the solar cell array capability. As mentioned previously, all these functions are carried out by means of the power conditioning and control systems (PCCS).

The PCCS can be classified into two main types of systems on the basis of their working principle, i.e., (a) dissipative systems, which do not extract maximum power from the solar cell array, and hence dissipate any unused power by employing shunt regulators, (b) non-dissipative systems, which extract the maximum power from the solar cell array employing optimum power tracker converter and hence they dissipate very little power internally. The bus voltage can be regulated or unregulated, irrespective of whether the PCCS uses dissipative or non-dissipative techniques. Thus, PCCS can also be grouped into, (i) regulated bus systems and (ii) unregulated bus systems.

3.1 Centralized Vs Decentralized

Whether the spacecraft power bus is regulated or unregulated, the spacecraft subsystems require different positive and negative voltages with varying regulation requirements. Therefore, the bus voltage is further regulated, levelled-up, levelled-down and/or inverted using regulators and dc-dc converters. If this process of further regulation, etc., is carried out at each load end separately, then such a concept is known as a decentralized regulation concept. On the other hand, if this process of further regulation is carried out in the main power system for all the loads, then such a concept is known as a centralized regulation concept. The decentralized regulation system has the advantage of being able to individually tailor bus power for each subsystem/load without any compromise. In a centralized regulation system, regulated voltages are determined as a compromise for all subsystems. Often it is found that the subsystems still need additional regulators and dc-dc converters for achieving further regulation and less ripple.

Depending upon the application, after a trade-off, optimum PCCS concept and system has to be selected. As an example, if the spacecraft is large, a lengthy harness is needed to interconnect the subsystems. The longer the harness, the larger the voltage drop in the harness and poorer the regulation if a regulated bus concept is utilized. The length of the harness depends upon the relative locations of subsystems or the payloads with respect to power system. In addition, proper isolation is very essential. To achieve this isolation, each user has to use either a dc-dc converter or a dc-ac inverter at their input. Thus, for this example, the decentralized regulation concept utilizing a non-dissipative unregulated bus approach may be the optimum choice.

3.2 Battery Charge-Discharge Control

The charge-discharge control unit continuously monitors the array output voltage and takes decision regarding to turning on/off of the battery and charging/trickle charging the battery. This control unit ensures that during the sunlit period of the orbit, the battery is charged in the predetermined mode (optimum tracker mode or constant current mode, etc.). When the battery reaches its maximum charge voltage (which varies as a function of temperature), it is trickle charged, typically at C/50 rate. The control unit also includes the protective features that will prevent the batteries under any circumstances from discharging below a predetermined level.

3.3 Solar Array Drive Unit

In three axis stabilized satellites (non-spinners), the solar panels should always face the sun to get maximum output. Hence, a drive unit is needed to drive the solar panels such that they are normal to sun all the time.

The solar array drive consists of (a) drive motor (stepping motor) (b) reduction gear assembly to transmit the drive motor power to the solar cell array (c) suitable slip-ring assembly to transfer power via the shaft to the satellite body (d) shaft encoder to indicate relative position between the solar cell array and spacecraft body, etc.

4.0 Design Considerations

For any spacecraft, the power requirements can be specified as, (a) Maximum power requirements with all subsystems functioning, and (b) Minimum power requirement to maintain altitude control, station keeping, telemetry and telecommand during eclipse. (a) dictates the size of the solar cell array while the capacity of the battery is determined by (b). In addition, additional allowance must be provided for meeting launch-time demands if it is found that the battery capability as specified by eclipse operation, could not meet the launch requirements. Alternately, a separate battery could be employed for launch-time demands. However if the spacecraft is launched by the shuttle, the launch requirements, to some extent, are provided by the shuttle. Hence the storage battery need not provide the launch power requirements. However, parking orbit phase and transfer orbit phase power requirements have to be considered in the design.

4.1 System Considerations

The configuration of the power system to yield optimum performance depends upon the satellite mission, anticipated active life, altitude of the orbit, inclination of the orbit, etc. Nowadays the satellite life expectancy is in the range of 5 to 10 years. Most of the communication satellites are in the geosynchronous orbit (35800 km, zero degrees inclination).

4.2 Voltage Selection

Depending upon the needs of various subsystems and chosen power system configuration, solar cell array bus voltage is chosen giving some allowance for the conditioners and battery to operate well and solar cell interconnections are done such that this voltage becomes one of the co-ordinates of the maximum power point.

4.3 Power Budget

The power system has to continuously provide the power required by subsystems and payloads or experiments and shall have typically a 10% margin. For example, a communication satellite contains in addition to power system, travelling-wave tubes (TWTs), receiver-transmitter chain, telecommand and telemetry systems, attitude and orbit control systems, etc. The power requirements of a typical geosynchronous communication satellite are in the range of 1500 to 2500 watts.

4.4 Solar Cell Array Area

The efficiency of 14.5% mentioned earlier

refers to the single solar cell. But in calculating the solar cell array size, the following factors have to be considered, namely, degradation due to radiation and micro-meteorites, UV damage, assembly losses including inter-cell wiring losses, isolation diode losses, seasonal effects (variation in the sun angle as the solar cell arrays are single axis/gimble driven and seasonal solar intensity variation), thermal factors (degradation or losses due to thermal cycling, high temperature operation), cover glass particle loss, mismatch losses, calibration errors, fill factor, wiring loss, array configuration, random and contamination losses, ground measurement errors, slip-ring losses, etc. The solar cell array shall provide full output power under orbital conditions after being subjected to shadowing resulting attitude or spacecraft body and appendage obscuration of the sun.

Depending upon the PCCS configuration, the power is processed through one or more power conditioners before it is supplied to subsystems or loads or is used for charging the storage batteries. The efficiency of these power conditioners, though in the range of 90% to 95%, is to be considered in the design. Typically, for a communication satellite in geosynchronous orbit with a life of 10 years, the end of mission power output from the solar cell array is in the range of 8 to 10 watts per square foot. Thus if a communication satellite power requirement is 1500 watts, it requires approximately 190 square feet of solar cell array.

4.5 Solar Cell Array Design

Though deployable solar panel array system is less reliable comparatively to the body mounted arrays, the improvements in attitude control system and giving considerations to other factors like large power requirements, low cost, deployment and weight standpoints, flexible substrate design is recommended for the solar cell array, which could be folded accordian-style during launch. This folded array could be sandwiched with protective pads, restraining hardware and mounting and tensing members between two honeycomb vanes that will hold tightly against the satellite body during launch by several restraining mechanism pads keeping opposing cover glasses from touching during storage. Normally, the total solar cell array area is distributed over two panels to maintain equilibrium.

4.6 Battery Capacity

As mentioned above, the duration and the number of eclipses depend upon the orbit altitude and its inclination. For a satellite in a geosynchronous altitude for 10 years of active life, the battery has to withstand 900 cycles of charge-discharge of 22.8/1.2 hours. Only essential functions of housekeeping and attitude control are operated during eclipse period. In addition, launch power has to be supplied by the battery (assuming that the spacecraft is launched by a rocket).

In view of weight optimization, it is better to use two batteries, one for on-orbit phase having large cycle life like Ni-Cd or Ni-H2 and an additional one shot (short life) but high energy density battery for launch needs exclusively or along with on-orbit battery.

Reverse voltage limiters are used across each cell in the battery string to improve the performance of the battery. When the solar bus voltage is above the preset level, the battery charge logic connects the battery to get charge. Battery charging is controlled by a combination of sensing of battery string voltage, battery temperature and individual cell pressure. For efficient and reliable operation of the battery, it shall be maintained between zero degrees C and 10 degrees C.

5.0 Conclusions

Spacecraft power systems consisting of solar cell arrays, storage batteries and PCCS have been presented in detail. The main differences between power systems for low earth orbit satellites and geosynchronous orbit satellites are also presented. Finally, important power system design considerations are included.

References

1. P.R.K. Chetty, "*Spacecraft Power Systems— Some New Techniques for Performance Improvement,*" Ph.D Thesis, Indian Institute of Science (I.I.Sc), India, 1978.

IMPROVED POWER CONDITIONING UNIT FOR REGULATED BUS SPACECRAFT POWER SYSTEM

Introduction

The regulated bus concept of power conditioning and control system has several advantages over the unregulated bus concept for geostationary spacecraft applications, namely,

a) Some loads may run directly from the bus, thereby improving the overall system efficiency;
b) Lighter load regulator/converter units are facilitated;
c) Low bus impedance is realized;
d) Solar array operating point is properly fixed.

Therefore, an optimization of design of this power conditioning unit (PCU) is highly beneficial. The regulated bus concept has, however, two main drawbacks. These are

a) Excess solar array power has to be totally or partially dissipated in a shunt regulator; this affects the thermal and mechanical design of the power system, and
b) Three types of regulators are required to control the power flow to and from the bus during the various operating modes encountered; they are, shunt, charge, and discharge modes. These constraints lead to a complex bus voltage control. These problems become more severe when redundancy is to be incorporated.

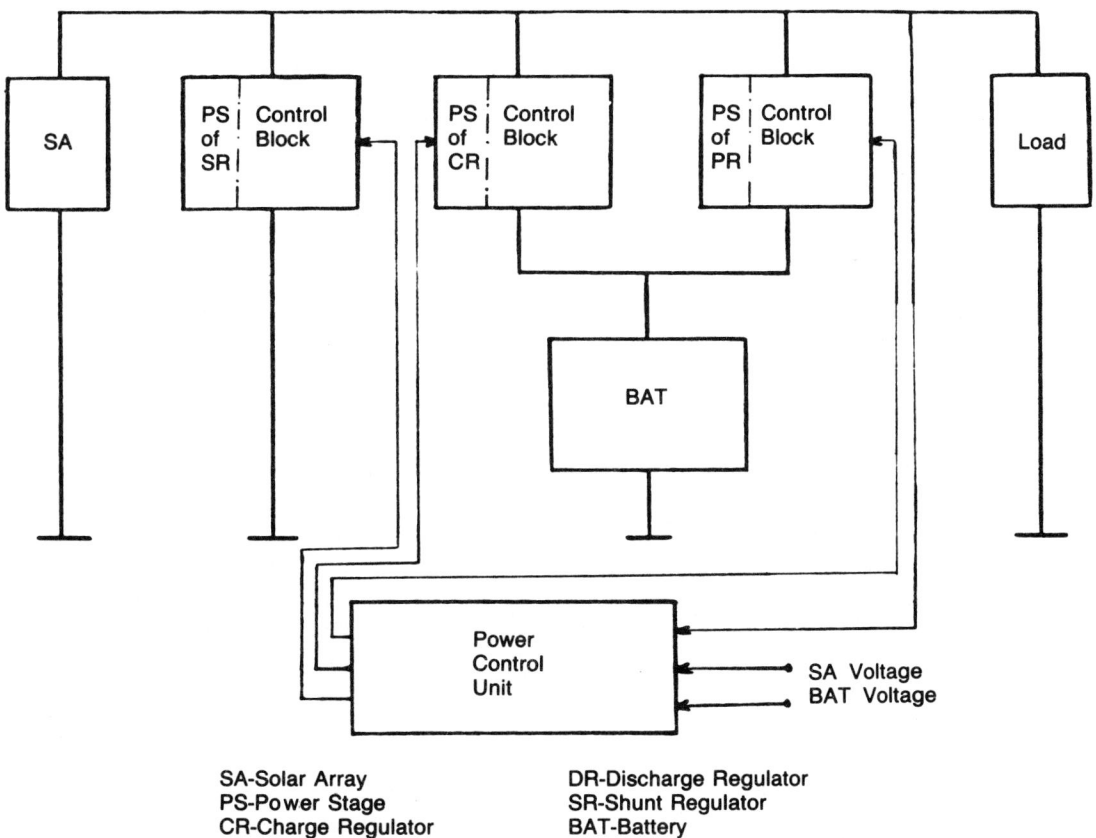

Fig. 2. Block schematic of conventional power conditioning unit for regulated bus power system.

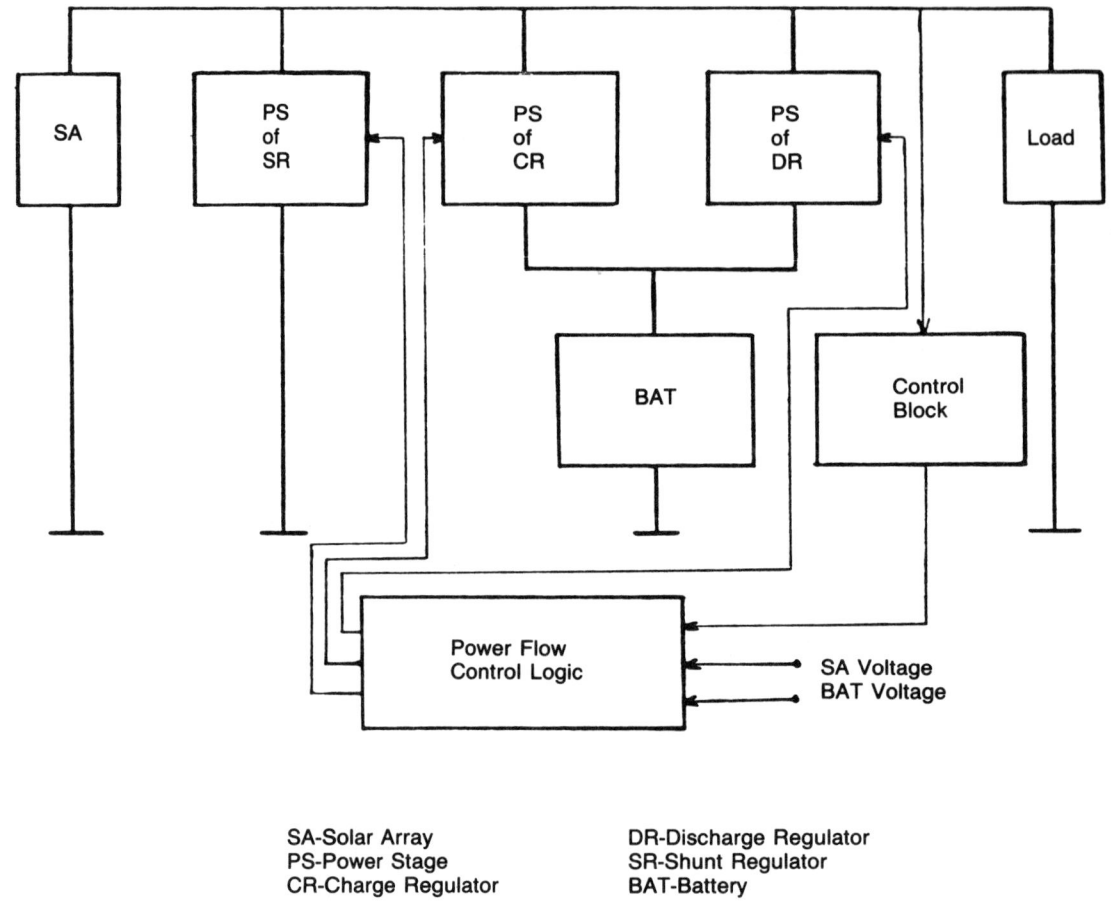

Fig. 3. Block schematic of improved power conditioning unit for regulated bus power system.

SA-Solar Array
PS-Power Stage
CR-Charge Regulator
DR-Discharge Regulator
SR-Shunt Regulator
BAT-Battery

A design approach to PCU is presented primarily to minimize the effects of the above two drawbacks. Figure 1 shows the block schematic of a conventional PCU, both the shunt regulator and the charge regulator work simultaneously. The shunt regulator maintains the bus voltage at a fixed value while the charge regulator charges the battery at a constant current. The shunt regulator, the charge regulator, and the discharge regulator have individual control blocks. The power flow control logic controls the regulators that are to be turned on/off depending upon voltage levels of the solar array and the battery.

In the new approach, a common control block, consisting of a reference voltage source, error amplifier, attenuator, and pulse-width modulator, drives the power stages of shunt, charge and discharge regulators as shown in Fig. 2. The charge regulator itself maintains the bus at a fixed voltage and charges the battery at a variable current. When the battery is completely charged, the shunt regulator is turned on and the charge regulator is turned off. A single common control block is employed to control the power stages of all the regulators instead of the individual control blocks in the conventional system. This is possible because at any instant of time, only one regulator is adequate to maintain the bus at a fixed voltage and/or charge the battery or supply power to the loads. In addition, as shown in Fig. 2, some components are made common for either of the regulators i.e.,

a) the inductor, which is usually heavy, is made common for the charge and discharge regulators;

b) the output capacitor, which is usually bulky, is made common for all the three regulators, as the complete unit can be mounted on a single printed-circuit board.

Theory of Improved Power Conditioning Unit

The operation of a single control block in three modes (shunt, charge, and discharge) is described first and that of the total unit is subsequently presented.

Shunt Mode Operation

Figure 3 shows the block schematic of the shunt bus regulator, which maintains the bus at a fixed voltage (V_o). The control is achieved by comparing the reduced bus voltage ($K \cdot V_o$) to a reference voltage (V_{ref}). The error voltage is amplified and fed to a pulse-width modulation (PWM) unit through an attenuator. The PWM unit is synchronized at a clock frequency, f, and its output is a pulse width τ proportional to the amplified error voltage but limited by the attenuator. This signal is employed to switch the transistor (Q1) into saturation or cut-off. Thus, the emitter current of Q1 (i.e., through R) is controlled and the bus voltage is regulated. When the shunt transistor is on, the diode D1 protects the filter capacitor (C2) from discharging through Q1 so that it provides the required current to the load. The relation between required output (P_o) and the power available from the solar array (P) at the bus voltage V_o, is given by

$$P_o = [V_o \cdot I_o] = (1/T) \int_\tau^T P \cdot dt \quad (1)$$

Extracted from Ph.D. Thesis of P.R.K. Chetty, entitled *"Spacecraft Power Systems—Some New Techniques for Performance Improvement,"* Indian Institute of Science (I.I.Sc), India, 1978.

where, I_o is the load current and T is the oscillation period ($= 1/f$). It is well known that P is dependent on the illumination and temperature conditions of the spacecraft. From the equivalent circuit of the solar array, the current I available from the array is given by

$$I = I_g - I_S \left\{ exp \frac{q(V + I \cdot R_s)}{(A \cdot K \cdot T_o)} - 1 \right\} \quad (2)$$

when the effect of shunt resistance is neglected. This equation can be rewritten as,

$$V = \frac{A \cdot K \cdot T_o}{q} \ln \left\{ \frac{I_g - I}{I_s} + 1 \right\} - (I \cdot R_s) \quad (3)$$

Neglecting the voltage drop across D1 (Fig. 3), the array output voltage is given by,

$$V_o = \frac{A \cdot K \cdot T_o}{q} \ln \left\{ \frac{I_g - I}{I_s} + 1 \right\} - (I \cdot R_s) \quad (4)$$

and

$$P = V_o \cdot I = \frac{I \cdot A \cdot K \cdot T_o}{q} \ln \left\{ \frac{I_g - I}{I_s} + 1 \right\} - I^2 \cdot R_s \quad (5)$$

combining eq(1) and eq(5)

$$P_o = (1/T) \int_\tau^T I \left[\frac{A \cdot K \cdot T_o}{q} \ln \left\{ \frac{I_g - I}{I_s} + 1 \right\} - I \cdot R_s \right] dt \quad (6)$$

Simplification of eq(6) leads to,

$$P_o = V_o I_o = I \left[\frac{A \cdot K \cdot T_o}{q} \ln \left\{ \frac{I_g - I}{I_s} + 1 \right\} - I \cdot R_s \right] \frac{T - \tau}{T} \quad (7)$$

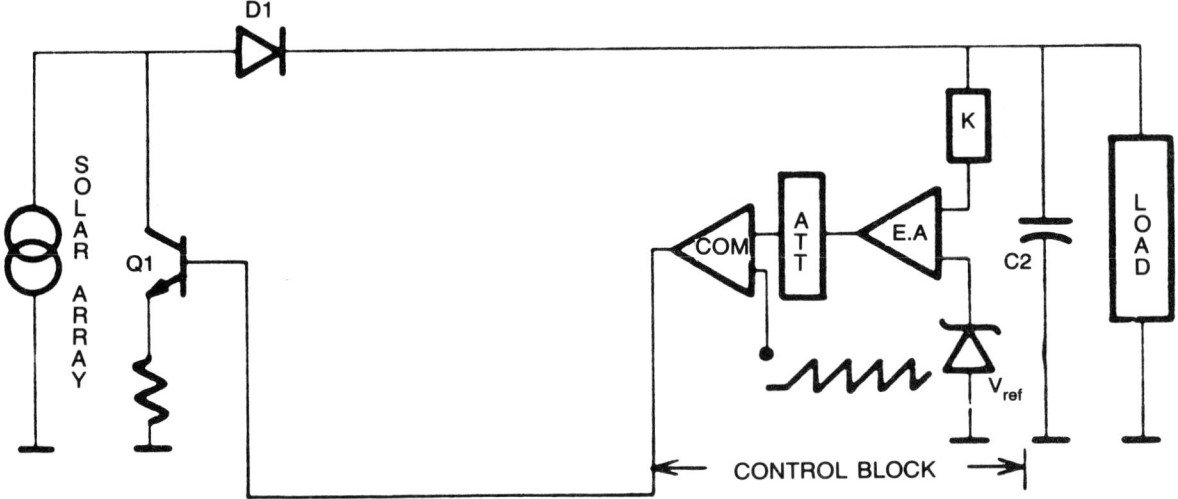

Fig. 4. Shunt mode operation of power conditioning unit.

From this equation, an expression for V_o is obtained

$$V_o = (1 - \frac{\tau}{T}) \left(\frac{I}{I_o}\right) \left[\frac{A \cdot K \cdot T_o}{q}\right]$$
$$\ln\left\{\frac{I_g - I}{I_s} + 1\right\} - I \cdot R_s \qquad (8)$$

where τ/T is the duty cycle of the PWM signal. It is clear from this equation that V_o varies as a function of the duty cycle of the PWM output signal.

Charge Mode Operation

The charge regulator is meant for charging the battery at a variable current while maintaining the bus voltage (regulated) at V_o. Figure 4 shows the block schematic of the charge regulator. The PWM signal, whose pulse width τ is proportional to the error voltage, is generated as in the case of shunt-mode operation This signal is employed to switch Q2 into saturation or cut-off; thereby controlling the current flow into the battery (B) such that the bus voltage is maintained at V_o. When Q2 is on, energy is stored in the inductor (L) and this main-

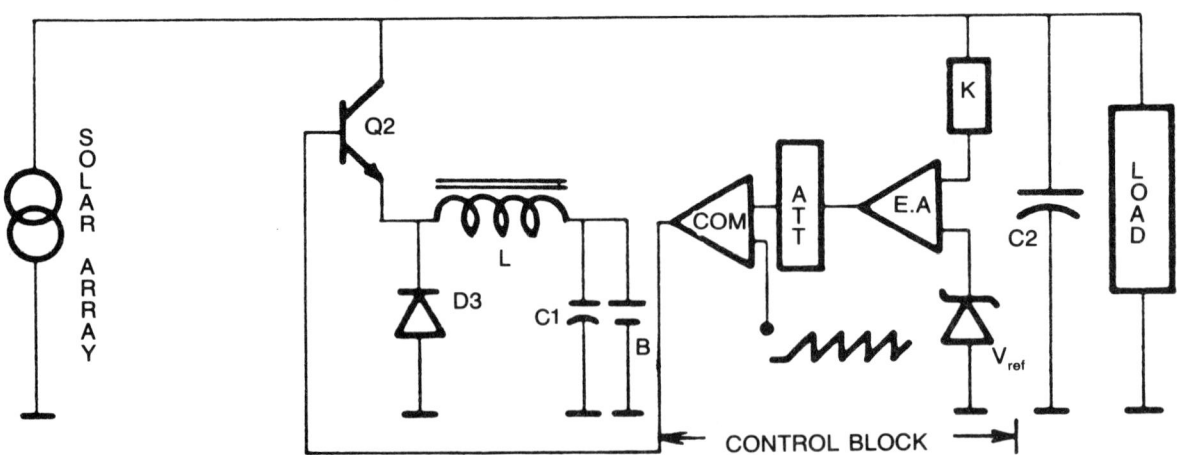

Fig. 5. Charge mode operation of power conditioning unit.

tains a continuous current flow into the battery when Q2 is off. The power input to the battery (P_b) and the pulse current (i_c) in L are related through the following equation

$$P_b = [1/T] \int_0^\tau \{V_o \cdot i_c\} dt \quad (9)$$

and i_c is given by

$$i_c = \left[\frac{V_o - V_b}{L} \right] t \quad (10)$$

where V_b is the battery voltage. Combining eqs. (9) and (10), followed by integration leads to,

$$P_b = V_b \cdot i_b = \frac{V_o(V_o - V_b)}{2T \cdot L} \tau^2 \quad (11)$$

From this equation, the expression for V_o is written as

$$V_o = \frac{V_b}{2} \left[1 + \sqrt{1 + \frac{8L \cdot i_b}{T \cdot V_b} \left(\frac{T}{\tau}\right)^2} \right] \quad (12)$$

Thus, V_o varies as a function of the duty cycle of the PWM output signal. If the pulse width is limited to a maximum value, in turn the duty cycle is also limited to a maximum value. As an example, if $\tau/T < 1$, then the battery is charged at a variable current i_b so that,

$$0 < i_b < i_{bm} \quad (13)$$

with i_{bm} as the maximum current given by the following equation

$$i_{bm} = \frac{V_o \cdot T}{2L \cdot V_b} (V_o - V_b) \left(\frac{\tau}{T}\right)^2 \quad (14)$$

It is easy to see from this equation that the charge current decreases as the battery gets charged up. When the battery is completely charged, the signal from the end of charge monitor unit is used to trickle charge the battery through an alternate path, e.g., by the resistor and diode connected across the collector and the emitter of the transistor Q2 (not shown in Fig. 4). At this stage, the shunt regulator is switched on, which maintains the bus voltage regulated at a fixed value.

The advantages of this method of smooth decrease of charge current at a rate governed by eq (14) avoids gas formation in the battery, which in turn improves the lifetime of the battery. This also offers a continuous transition between the extreme states of the battery. Thus, the charge regulator works in a similar way to the shunt regulator, employing all the available power for charging the battery, instead of dissipating in a shunt regulator as in the case of the conventional regulated bus system. This improves the overall efficiency of the system.

Discharge Mode Operation

Figure 5 shows the block schematic of the discharge regulator. This is boost-type switching regulator, which boosts the battery voltage to the bus voltage level and maintains it. The control is achieved by the PWM signal, whose pulse width is proportional to the error voltage, generated as in the earlier case. This PWM signal is employed to switch the Q3 into saturation or cut-off. When Q3 is on, energy is stored in the inductor (L) which is in series with the battery. When Q3 is off, the stored energy in L tries to collapse, reversing its

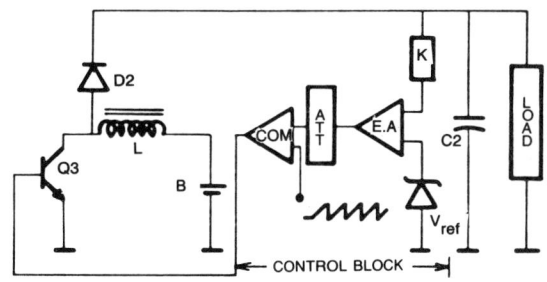

Fig. 6. Discharge mode operation of power conditioning unit.

polarity. Thus, the output capacitor is charged to the bus voltage level and is maintained at V_o, which is higher than the battery voltage. The operation of this circuit is briefly analyzed below.

When Q3 is on, energy is stored in L and V_b is the source voltage that gives out the energy.

$$i = [V_b \cdot t]/L \text{ between } 0 \text{ and } \tau \quad (15)$$

as inductor current increases linearly with time. When Q3 is off, the current which is at τ decreases gradually and again the battery is the source through which the current flows. The stored energy in L is released with decreasing i between τ and T, with a voltage of V_o-V_b. But during the interval (T-τ), L gives away energy which is stored during the period O-τ and V_b again supplies power at a current of i between τ and T as given below:

$$i = \frac{V_b}{L} \left(\frac{\tau}{T-\tau}\right)(T-t) \text{ between } \tau \text{ and } T$$

Hence, the power output P from the battery is given by

$$P = [1/T]\left[V_b \int_O^\tau i(t)dt + V_b \int_\tau^T i(t)dt\right] \quad (17)$$

Combining eqs. (15) and (16) into eq (17), and its integration leads to,

$$P = V_o \cdot I_o = \frac{1}{2T \cdot L}\left[\frac{V_b^2 \cdot V_o}{V_o - V_b}\right]\tau^2 \quad (18)$$

as

$$T = \tau \cdot V_o/(V_o - V_b) \quad (19)$$

From this equation, an expression for V_o is written as

$$V_o = [T/2L][\tau^2/T]\frac{V_b^2}{I_o} + V_b \quad (20)$$

Thus, V_o varies as a function of the duty cycle τ/T of the output signal of the PWM unit.

Complete Power Conditioning Unit

From the above discussions, it is clear that the control block is required in all the three cases. As only one of the regulators work at any instant of time, a single control block is used for all the three regulators in this scheme, instead of having three separate control blocks as in the case of a conventional power conditioning unit. The heavier elements like the inductor is used as a common element for the charge regulator and discharge regulator and the bulky output capacitor as a common element for all the regulators. While the common control unit is used for driving the power stages of charge, discharge, and shunt regulators, a simple power flow control logic routes the driving signal to the proper regulator. Figure 6 shows the block schematic of the complete conditioning unit.

Power Flow Control Logic

In any spacecraft power system, when the spacecraft is in the sunlit portion, the solar array is used to provide power to the loads and for charging the battery. When the battery is completely charged, then the battery is put into trickle charge and the shunt regulator is turned on which maintains the bus at a fixed voltage. When the spacecraft enters into the eclipse portion of the orbit, then the stored energy from the battery is used to provide power to loads. Also when the solar array is not capable of supplying the peak load requirements, its voltage will go down, which is used to turn on the discharge regulator. In this mode, both the solar array and the battery share to meet the load demand. Power flow control logic unit is required to facilitate all these functions.

Design

As described earlier, the regulated bus concept of power conditioning and control system has several advantages. This unit has been designed to the following specifications:

Solar array open circuit voltage   40-50

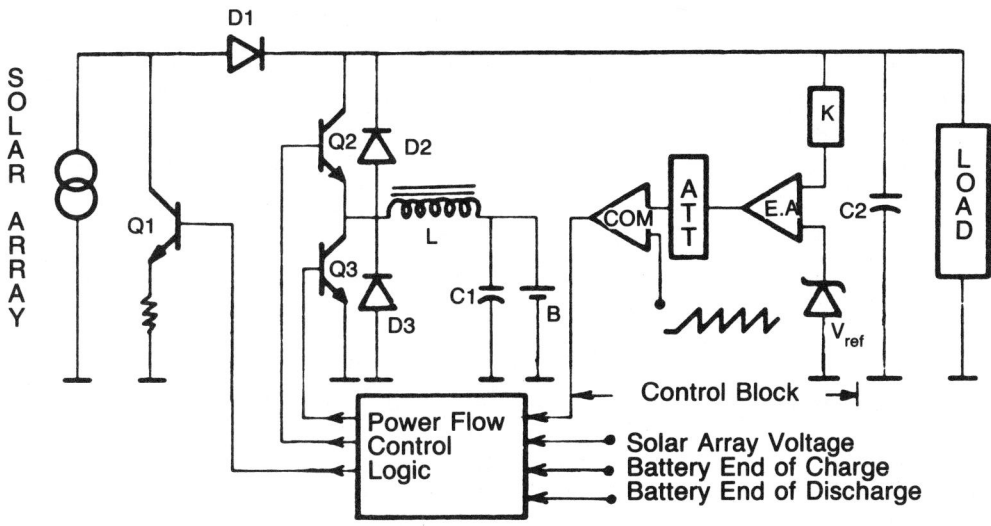

Fig. 7. Block schematic of power conditioning unit.

Solar array current at bus voltage	1 A ± 0.1 A
Bus voltage (V_o)	30.5 ± 2% Vdc
Bus regulation	± 1%
Bus Ripple	Less than 250 mV at full load current.
Continuous load current	0.5 A at bus voltage

The remaining power of the solar array is used to charge the battery that provides the power to the load during orbital eclipse period.

Experimental Results

The improved power conditioning unit has been breadboarded and tested. This unit exhibited excellent performance, which is briefed below:

Performance at the full load current of 0.5 A

a) When the charge regulator is working,

    Bus Voltage = 30.7 V
    Bus Ripple = 140 mV

b) When shunt regulator is working,
    Bus Voltage = 30.5V
    Bus Ripple = 200 mV

c) When discharge regulator is working,

    Bus voltage = 30.8 V
    Bus Ripple = 200 mV

Operating switching frequency = 12.2 kHz

It is easy to see that the performance of the unit closely follows the specifications of the unit. Hardware comparison of the improved power conditioning unit and the conventional power conditioning unit is given in Table 1.

Conclusions

Although the power conditioning unit described here has been designed for a power rating of 30 W, the same can be extended to the kilowatts range by increasing the rating of the power stage and/or by multiphase operation. In the latter case, more of similar modules can be added in parallel. This also reduces the filtering problems as the effective frequency of operation is increased. As the number of components used is comparatively small, redundancy may be incorporated without adding too much weight for the system. As all the blocks employed in the unit described above are available in the modular form, modularization is easy. Besides, the common use of subunits-components

Table 1. Hardware Comparison of Conventional and Improved Power Conditioning Units.

Parameter/Unit	Conventional Unit	Improved Unit
Control Stage		
1. Voltage divider network for sensing the bus and/or sensing the voltage proportional to charging current	3	1
2. Voltage references	3	1
3. Error comparator Amplifier	3	1
4. Pulse width Modulators	3	1
Power stage		
5. Inductors	2	1
6. Capacitors (High Value)	3	1

reduces the soldering and cabling problems and hence the weight is reduced. All these factors improve the reliability of the improved power conditioning unit.

References

1. P.R.K Chetty, *"Spacecraft Power Systems—Some New Techniques for Performance Improvement,"* Ph.D. Thesis, Indian Institute of Science (I.I.Sc), India, 1978.

Chapter 8

Reliability

Design for Reliability 160

Reliability and Redundancy 167

Reliability and Failure Mode and Effects Analysis 170

DESIGN FOR RELIABILITY

1.0 Introduction

Reliability is the main requirement of any system or equipment. It is defined as the probability of performing a specified function under specified conditions for a specified time. For applications, where the system or equipment becomes expensive and that there is no possibility of repair as in the case of satellites, the necessity of built-in reliability becomes a rule rather than exception. Therefore continuous efforts are being made to realize better and better and higher reliability of systems or equipment used in space.

There are many ways for improvement of the reliability of any electronic equipment or system, the most important of which include the component improvement, stress-level reduction, circuit or system design simplification, conservative design, and the selection of proper components and manufacturing techniques. Only after these methods have been fully exploited to increase inherent circuit or system reliability, should redundancy be considered to take care of chance failures. When judiciously employed, higher levels of reliability can be accomplished by redundancy either on a piece part or circuit basis.

Thus, the following sections deal with various methods or techniques to be considered in designing the systems or circuits for reliability.

2.0 Reliable Circuit Design

After considering the individual component reliability, namely, improvement by proper derating, its interaction with supply voltages, its mechanical arrangements and the environment under which it has to operate, etc., then the designer shall examine the circuit and assess its reliability. In fact, the circuit design is another aspect of the environmental design as the circuit is the immediate electrical environment of each component. It is convenient to deal with circuit reliability separately because of the wide variety of requirements that the circuit design must meet.

First, the circuit must fulfill its function; secondly, the design of the circuit has a bearing on the reliability of its components. Following the initial design of an electronic system, a mathematical analysis must be made of the individual circuits to determine whether they will permit the initially set system reliability goal to be met. Good design is universally regarded as a quality that can be equated to good reliability.

The three factors that contribute to the reliability of a system or equipment as it leaves the manufacturer are the reliability of the design, the reliability of the component parts employed, and the reliability involved in the manner in which the components and the design are fabricated. The reliability of these factors, each of which is equally important, is expressed in the following equation.

$$P_a = [P_d \cdot P_c \cdot P_f]$$

where P_a = the overall reliability
P_d = the design reliability
P_c = the component reliability
P_f = the fabrication reliability

3.0 Design Reliability

As explained above, the design reliability is equally important as the component or fabrication reliability. Although the component and fabrication reliability was very good, in the case of a regulator, because of its poor stability in a particular operating mode, the circuit did not deliver the expected output. Thus, design reliability is equally important and is dealt with some illustrative examples.

3.1 Power Conditioning and Control System

Sometimes a system level approach is needed in the case of a large system, like a satellite, to improve the reliability, efficiency, and to avoid unnecessary waste of power. It is no wonder if one finds several voltage regulators in series between the source of raw power and the user. Each of these regulators has inefficiency and unreliability and

---

Portions are reprinted with permission from *Electronic Engineering*, UK, Aug 1979.

hence much of the available raw power is wasted as heat and the reliability of the overall system has decreased. Such a situation arises when individual units are developed in isolation and can be avoided with good control at the system level.

3.1.1 Common Regulator

When a single regulator powers two or more subsystems and if there is no provision to cut off power to each individual subsystem, in such cases to avoid the propagation of the malfunction at one such subsystem to other subsystems, a fault isolation circuit (FIC) is placed between such subsystem load and common regulator as shown in Fig. 1. In such cases if the load fails in short circuit suddenly, the control circuit of FIC should respond faster and isolate the failed load; otherwise the common regulator might go into current limit or foldback condition temporarily cutting off the power to all loads. Thus FIC shall respond not only to any bad impedance but also to any rate of change of load impedance. Normally the series-pass transistor of FIC will be failing because of two reasons (a) any reverse voltage, which may be delivered by an active load, (b) any fault in any other user that short circuits the regulator for a short time, during which input voltage to FIC drops to zero. Then the voltage across the output capacitor of FIC is applied between the collector and emitter, first zener breakdown could have occurred and the transistor would probably have been destroyed before the fault is isolated. This problem can be avoided by connecting a diode across the series-pass transistor. This allows the output capacitor to discharge rapidly.

3.1.2 Input Filtering

For certain applications, L-C filters are required on power supply interfaces. In such cases, transient analysis becomes important during switch-on, both the current and voltage may overshoot to levels that could cause damage to components. Large negative voltage transients would occur at the power subsystem/user interface when the relay contact was opened (Fig. 2A), which can result in component damage. This can be avoided by adding a diode as shown in Fig. 2B.

3.2 Driving

The simplest drive source is a resistor with a pair of clipping diodes (Fig. 3A) or a voltage reference diode (Fig. 3B). If the driving source voltage levels are unsuited to either of these methods then a transistor driver circuit can be employed. Two such circuits are shown in Figs. 4A and 4B.

3.3 Inputs Driven from Switches

When a gate is driven by a mechanical contact, the circuit arrangement shown in Fig. 5A may be employed. The resistor reduces the possibility of noise pick-up when the switch is open. However,

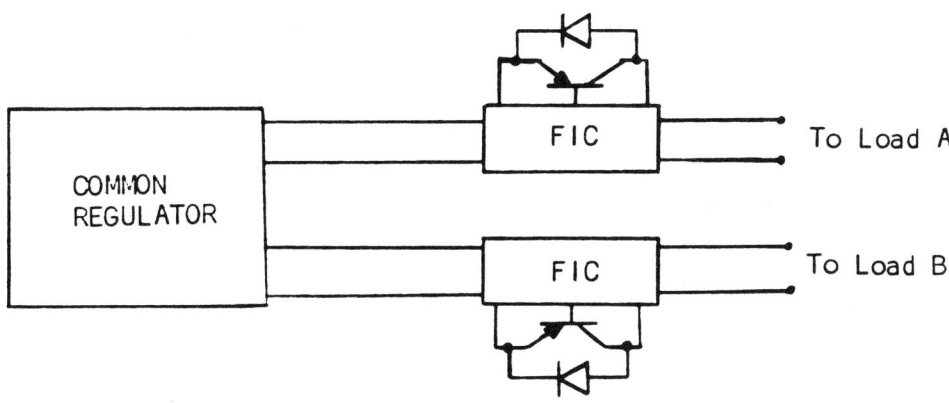

Fig. 1. Input short circuit protection for fault isolation circuit.

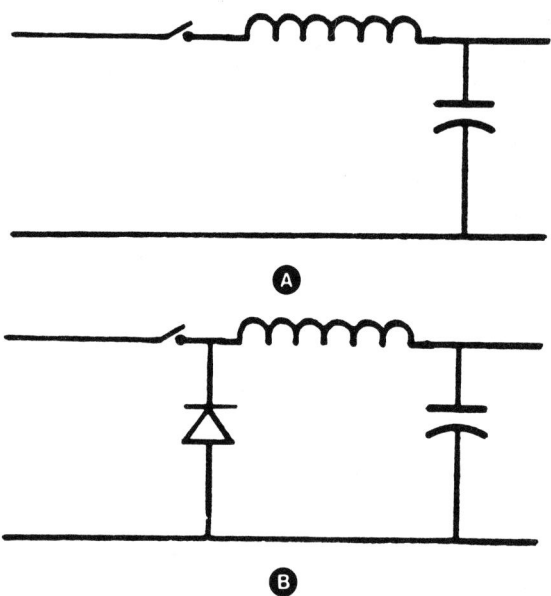

Fig. 2. Protection of LC filters in power supply interfaces.

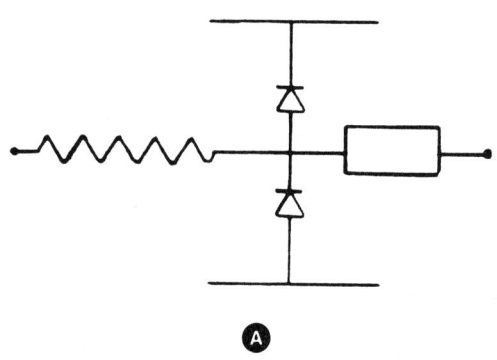

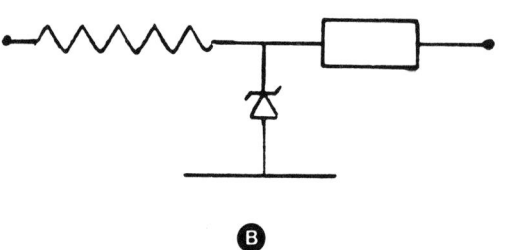

Fig. 3. Resistive source (A) with clipping diodes, (B) with voltage reference diode.

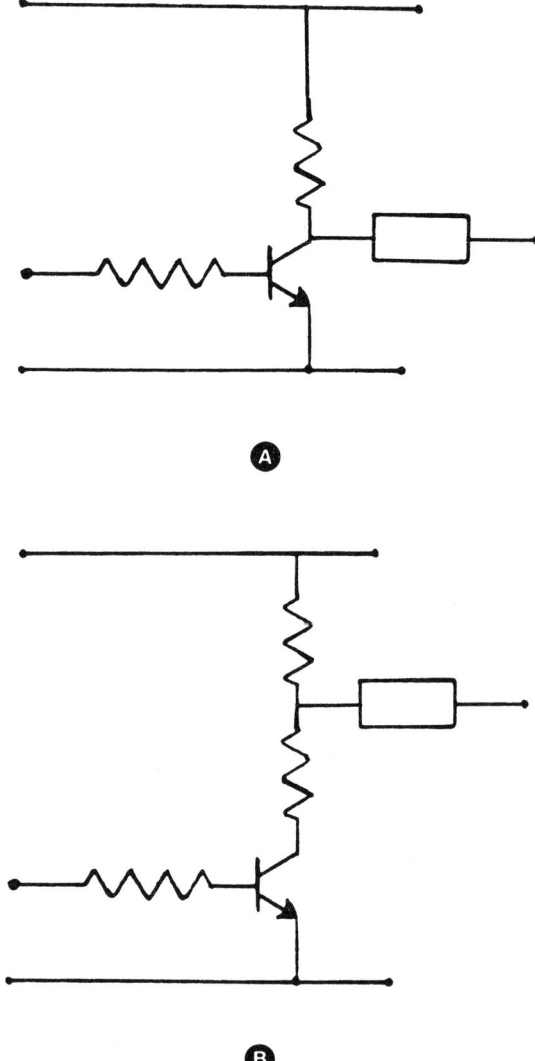

Fig. 4. Drive source interfaces.

contact bounce in mechanical switches causes the generation of a train of spurious pulses at each operation. It is better to use the switch to trigger a monostable, which can provide a single pulse if required, or delay the application of the signal until switch bounce is ended. Alternately, if the changeover switch is available then a pair of cross-coupled gates may be used as shown in Fig. 5B. Also bounce-free buffer gates can be employed for such applications.

immunity. Sometimes an RC network tends to integrate any repetitive waveforms at the input. Depending upon the mark, space ratios of the noise, and the relative charge and discharge time constants, the voltage on the capacitor may build up and eventually reach a sufficient level to make the transistor conduct.

A simple circuit shown in Fig. 6B [5] overcomes all of these drawbacks. This circuit discriminates the command signal by its amplitude as well as its width and has high noise immunity because of its hysteresis.

3.5 Digital Interface

In the case of data commands, normally 8 bits or 16 bits of information is transferred to a user sub-

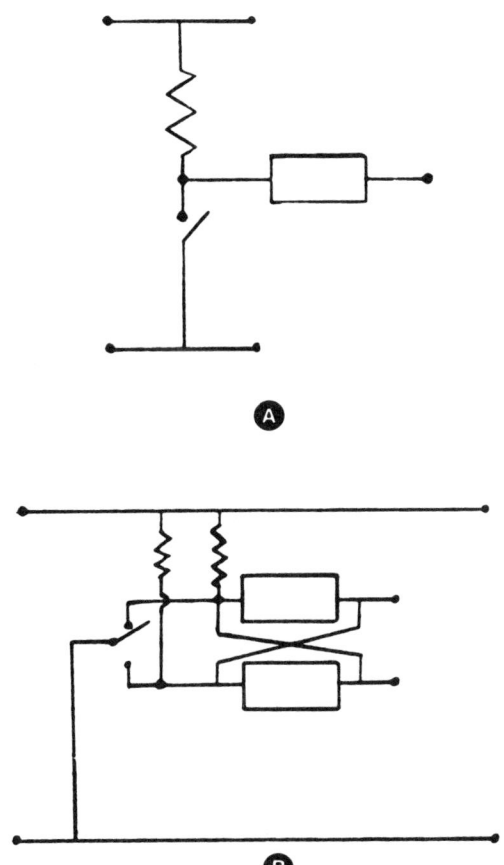

Fig. 5. TTL gate driven by mechanical contact (A) directly, (B) using cross coupled gates.

3.4 Interface with Relays

Relays and relay drivers are the peak power consuming circuits operating for short duration on command. Most often they respond to noise, if the immunity of the circuit is not good. So foolproof operation of relay drivers plays an important role in the successful operation of a system or equipment.

To descriminate between a true command and noise, the user interface circuit should make use of the distinct amplitude and width of the command pulse. A commonly used command interface employs an RC filter at the input (Fig. 6A). This circuit lacks hysteresis, which provides the noise

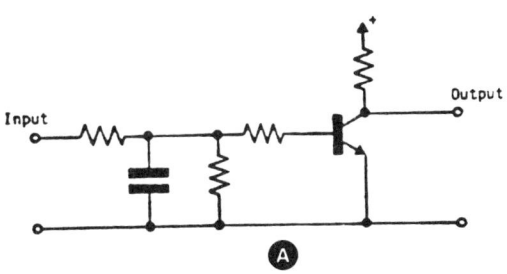

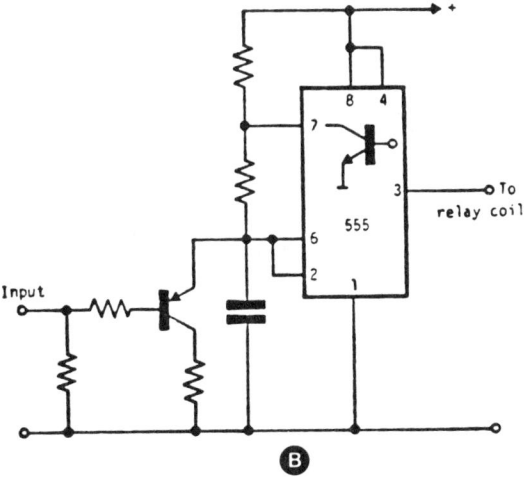

Fig. 6. Relay driver circuits.

163

system. A data command user receives a clock line, a serial data line, and an address line. Each user is allocated a separate address and therefore requires a decoding circuit to recognize when the transfer of serial data is intended for him. When the correct address is decoded, a simple pulse is generated to enable the clock and data inputs to the shift register. The output of such shift register will be random when the power is applied or restored following a temporary failure, which can cause a malfunction. For example, random output of such register can turn-on/off critical relays. For example, in a spacecraft this could result in the loss of attitude with subsequent loss of solar array power, thermal problems, etc. This can be avoided by designing the circuit such that when power is applied or restored, the shift register outputs are set automatically to a predetermined desirable safe state. Until this setting is carried out, the execution of any commands can be deferred.

3.5.1 TTL Integrated Circuits

The design of reliable systems using ICs necessitates the application of certain precautions and special techniques to the functional design as well as to the power supply decoupling, etc. The unused inputs of all OR and NOR gates and of the unused sections of AND, OR, INVERT gates must be grounded. For the best noise immunity, clock rise and fall times should be appropriate. Preset and clear pulses should be present for longer than the clock pulses. Unused preset, clear, J, and K inputs should be treated in the same manner as are unused gate inputs.

3.6 Level Shifting

Many currently marketed COSMOS devices require the conversion of standard TTL logic levels to higher voltages. Level shifting circuits must have a low output impedance during the edge transition, to drive the capacitive COSMOS inputs. A high impedance output in one direction is possible in either the 'HIGH' or the 'LOW' state: for example, an npn emitter follower output will have a high output impedance for positive signals. Thus, if a signal of the

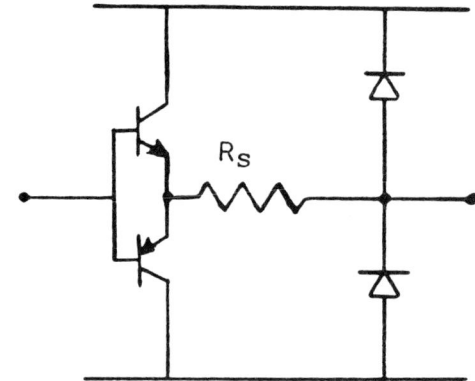

Fig. 7. Diode protection for COSMOS level shifting circuit.

correct polarity is induced at the output because of say, crosstalk in the load, then the driven COSMOS output may be taken out of its specified limits. A diode connected from the driver output to the power rail as shown in Fig. 7 will 'protect' the device and normally only one of the logic levels requires this protection. The resistance R_s in series with the output also shown in Fig. 7 has the effect of reducing any ringing in the lines that might be caused by the series combination of the printed-wiring inductance and the COSMOS output load and stray capacitances. Nevertheless, initial precautions against such ringing are recommended and the level-shifting circuits should be located as closely as possible to the COSMOS array they drive. Excessive ringing can seriously impair the operating margin of COSMOS devices.

3.7 Operating Conditions

Manufacturers' data sheets provide information about the operating conditions of the components and it is the responsibility of the designer to evaluate the conditions i.e., voltage, current, power dissipation, safe operating area (forward and reverse), temperature to which the component will be subjected and to ensure that in no case the limits are exceeded.

For an example, the rectifier-filter of the dc-dc converter secondary was designed assuming that the load will be connected permanently and no bleeder resistor has been used. This poses a con-

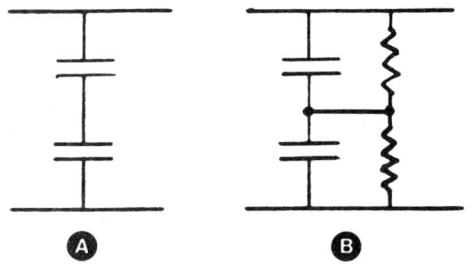

Fig. 8. Capacitor's protection when they are connected in series.

dition that the power supply shall not be turned-on when there is no load. It is noticed that sometimes the power supply is turned-on without any load being connected and output capacitors have failed as they are selected assuming that the load is always connected. When there is no load the output voltage builds up to a higher level than with a load.

Another example is when two capacitors are connected in series to meet the requirement of capacitors with high voltage ratings, proper care should be taken to avoid the failure of capacitors due to unequal potential division between the capacitors. Unequal potential division is possible because the tolerance of capacitors is of the order of 20%. The unequal potential division can easily be avoided by using two resistors in parallel with capacitors with five percent tolerance, as shown in Fig. 8.

3.8 Component Tolerances

The worst case design philosophy is commonly followed for designing reliability into an electronic circuit. A circuit design is to be checked for its performance as a function of component tolerances. The circuits that work well when they are new, may not work later because of aging. Some amplifiers may oscillate as the component values change. Hence, worst case design analysis has to be carried out taking into account the component tolerances.

4.0 Component Reliability

Component reliability plays an equally important role in the overall reliability of the system or equipment. The reliable performance of a particular component can be considerably improved by operating it at reduced stress levels. Typical derating factors are as follows: carbon composition (RCR) type resistor has to be derated to 50% of the rated power, maximum voltage shall not exceed 80% of the maximum rated voltage. In the case of digital IC's 80% of the manufacturer's maximum fan-out is allowed. In the case of linear ICs, the manufacturers suggested bias voltages shall not be derated unless precautions are taken to ensure that such action does not cause possible malfunction. Tantalum (wet) capacitors shall be derated to 50% of rated voltage. These are only some examples.

Also depending upon the application, namely the environmental and circuit conditions, the components have to be selected. For example, solid types should not be used in low impedance circuits (power supply filters) unless protected by a series current-limiting resistor.

5.0 Fabrication Reliability

Fabrication reliability is equally important as illustrated in some of the examples below. A circuit or system may operate satisfactorily and reliably when alone but may exhibit unexpected behavior when assembled or integrated into a complete system. The disturbance observed often effects not only the performance of the system, but also component operating conditions and therefore life. If fabrication is poor, then the system or equipment will not function as expected even though the design and component reliabilities are very high. Thus, equal consideration has to be given to the fabrication aspect of an equipment or a system.

5.1 Testing

To ensure reliability, the testing of systems has to be done at various stages of production. Besides the production of systems or equipment always involves a number of "do's" and "don'ts" which each manufacturer or contractor must adhere to. Certain materials create corrosion problems during long storage or use of which must be avoided. Certain procedures like soldering, if not conducted in the prescribed manner, may cause hardening of terminals and result in subsequent failures.

If the operation of an IC is in doubt, it is commonly removed from its socket and replaced by another, without first turning-off the power supply. In such a case, excessive inrush currents can occur and it is possible for a complete batch of new ICs to be destroyed one after another by an inadequately-briefed operator.

In some cases, the supply voltage to a module containing ICs can, by handling errors, be made higher than the rated voltages of the IC. Figure 9 shows an example of a supply arrangement where the LT supply voltage is dropped to a value appropriate to the IC by a series resistor and a decoupling capacitor. However, a hazard can arise if the IC is removed from the socket and replaced without switching-off the supply. A particularly severe hazard is represented by the charge potential developed across C1 while the socket is vacant, which will reach the full voltage. The safeguards are, (i) clearly place warning notices on the equipment and in the service manual, and (ii) solder the ICs directly to the panel.

5.2 Fabrication/Assembly

The reliability of any system is highly dependent upon the techniques used in the assembly of both the system and individual components. The effect here is very profound in that poor workmanship can result in poor reliability while good workmanship can be effective in improving the reliability. To cite an example of the impact of adopting good practices, tapped winding leads from toroids have been observed to lead to problems when both heavy gauge and fine gauge wires are wound over each other. The fine wire if wound over the heavy wire, has a tendency to sink down into the interstices of the large winding. Thermally induced movement of the large size wire will pinch the small size wire and cause open circuits. This is usually prevented by wrapping a layer of teflon tape over the heavy winding before installing the fine wire winding.

6.0 Redundancy

Though the inherent circuit reliability could be maximized as illustrated in the previous sections by reliable circuit design, judicious selection of components, proper derating and good workmanship, the chance failure that could partially or totally jeopardise a mission can be taken care of only by adopting proper redundancy to ensure the overall mission reliability at the required level.

Conclusions

Reliability is the important characteristic of any system or equipment. For applications, where the system or equipment becomes expensive and that there is no possibility of repair as in the case of satellites, the necessity of built-in reliability becomes a rule rather than exception. Thus, the various techniques for the improvement of the reliability with particular reference to the reliable circuit design are presented.

References

1. M.H. Dryden, "Design for Reliability," Mullard Technical Communication, No. 130, April 1976, pp. 395-432.
2. S.A. Meltzer, "Designing for Reliability," *IRE Transactions on RQC,* PGRQC-8, Sept 1956.
3. A.D. Web, "Designing Electrical Interfaces", *Proceedings of Spacecraft Power Conditioning Seminar,* 21-23 Sept 1977, ESA-SP-126, pp. 39-48.
4. Norman G. Dennies, "Insight into Standby Redundancy via Unreliability," *IEEE Transactions on Reliability,* Vol-R-23, No-5, Dec 1974, pp. 305-313.
5. P.R.K. Chetty and N.V. Sivaprasad, "An Improved Relay Driver Circuit," *Electronic Engineering,* U.K., Aug 1979.

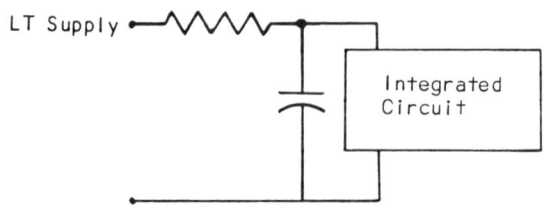

Fig. 9. High voltage hazards.

RELIABILITY AND REDUNDANCY

1.0 Introduction

Though the inherent circuit reliability could be maximized by optimized circuit design, judicious selection of components, and good workmanship, the chance failure that could partially or totally jeopardise a mission can be avoided only by adopting redundancy. When judiciously employed, higher levels of reliability can be accomplished by redundancy either on a piece part or circuit basis. This ensures the overall mission reliability at the required level. Thus various redundancy approaches are presented here.

2.0 Redundancy Approaches

There are different approaches to redundancy such as standby redundancy, load sharing redundancy, majority logic redundancy, shared mode of standby redundancy, partial redundancy, etc., which are described below in detail.

2.1 Standby Redundancy

In this case only one of the two or more similar power supplies is in operation at any instant of time. This configuration is shown in Fig. 1. two independent failure detectors are needed to monitor continuously the performance of the power supplies. Thus, the failure detectors are powered continuously to make the system more reliable. An alternative is to switch-on the particular failure detector of the working power supply only. In that case it is assumed that the failure detector and change-over network are highly reliable and the redundant units do not degrade while inoperative. An important point to be noted with this standby redundancy is the existence of a momentary power outage during change-over without special energy storage provision. This exists for a maximum period of 3 milliseconds, which is the relay contact change-over period. If this is not acceptable then standby redundancy should be excluded from such systems.

2.2 Load Sharing Redundancy

In this case, two or more power supplies operate at the same time, sharing the total load. If a power supply fails, that unit is isolated and the remaining power supplies continue to operate, sharing the load in larger proportions. This configuration is shown in Fig. 2. There will not be any power outage during change-over. In this type, the power supplies operate at other than the maximum efficiency design point. Therefore the total system efficiency is lower.

2.3 Majority Logic Redundancy

In this case, three or more power supplies operate at the same time and in parallel. The failed unit is isolated and the remaining units share the load in larger proportions. Though individual failure detectors are not required, an integrated failure detector is required to perform the comparison of functions. This configuration is shown in Fig. 3.

2.4 Shared Mode of Standby Redundancy

In this case, two loads A and B are supplied by

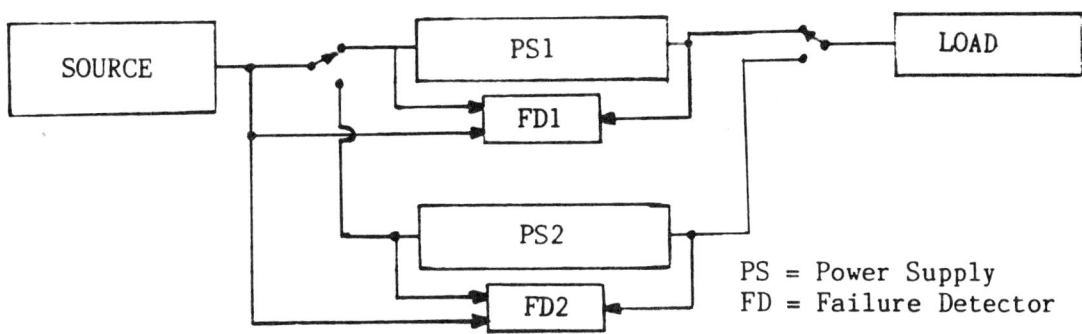

Fig. 1. Standby redundancy.

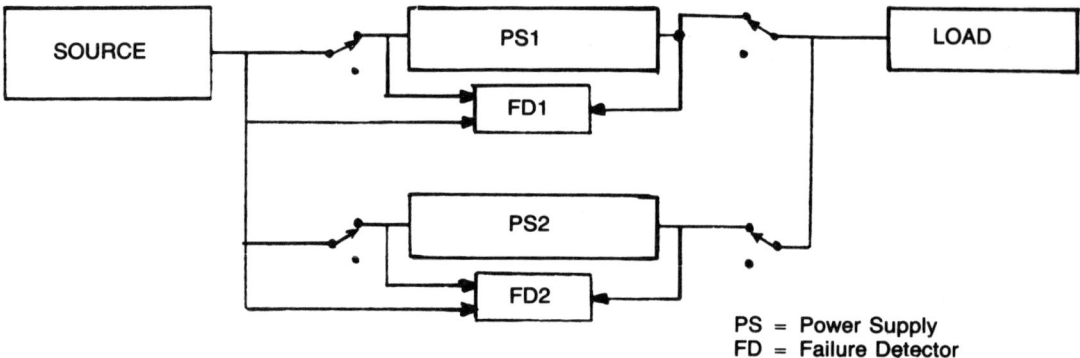

Fig. 2. Load sharing redundancy.

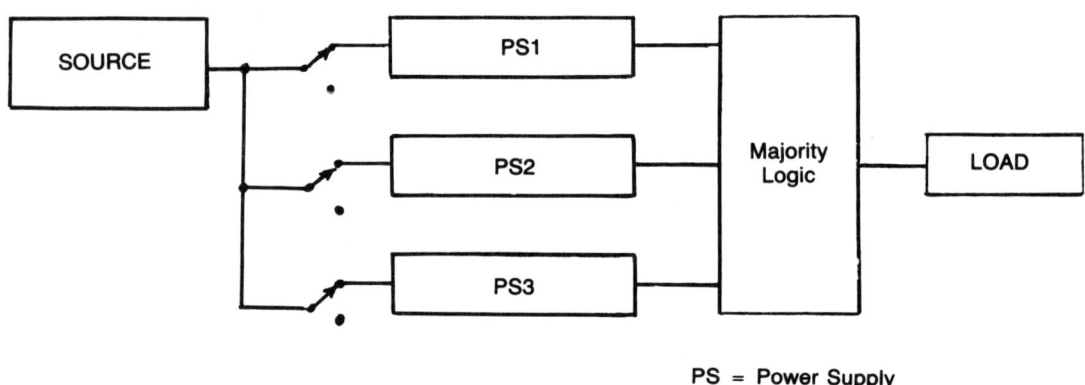

Fig. 3. Majority logic redundancy.

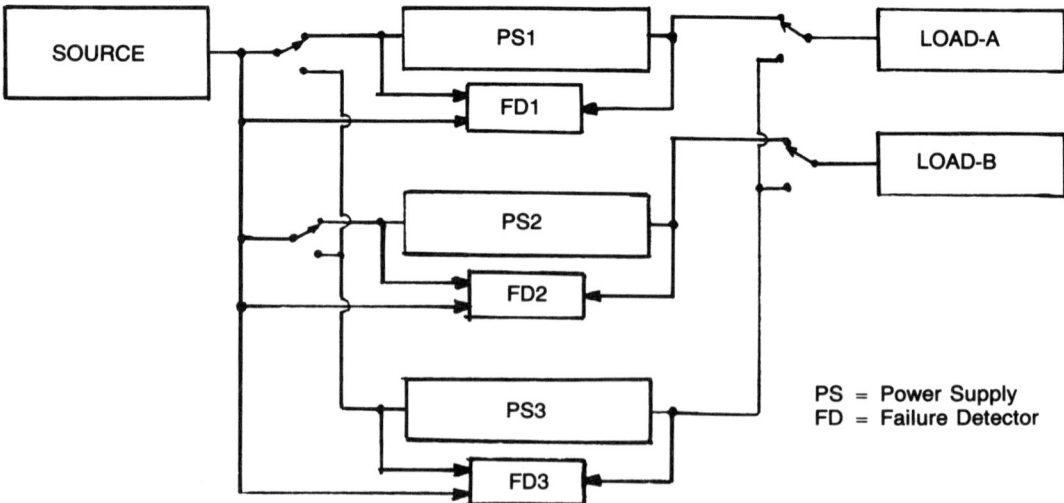

Fig. 4. Shared mode of standby redundancy.

two power supplies and there is only one common redundant power supply. Three separate failure detectors are employed. This configuration is shown in Fig. 4. The redundant power supply is capable of taking the sum of the loads A and B. In case of failure of power supply 1, redundant power supply provides power to the load A, isolating the power supply 1 from the source and load A. This system operates as shared mode of standby redundancy, initially working as a standby redundancy having a single redundant power supply for two power supplies 1 and 2. If a power supply fails then the redundant power supply works at lower efficiency as it is designed for the sum of the loads A and B. Power supply 2 together with the redundant power supply work as if they are sharing the load, though they are supplying independent loads. If the power supply B also fails, then the redundant unit itself supplies the load A and B.

2.5 Partial Redundancy

This type of redundancy, as shown in Fig. 5, is based on the fact that the failure rates of passive components are lesser than those of active components. Hence, instead of employing a complete circuit as a redundant unit, judiciously selected parts or sub-circuits primarily containing active components are only duplicated as redundant elements.

3.0 Optimization

In selecting a redundant approach, the following main factors shall be considered, i.e., reliability, weight, circuit complexity, efficiency, cost, etc. One has to select the redundancy approach depending upon the mission requirements and by carrying out a trade-off analysis giving appropriate weightages for different performance parameters of interest to that particular mission.

4.0 Failure Detector

The detection of failure at the first instant is very important and the failure detector shall con-

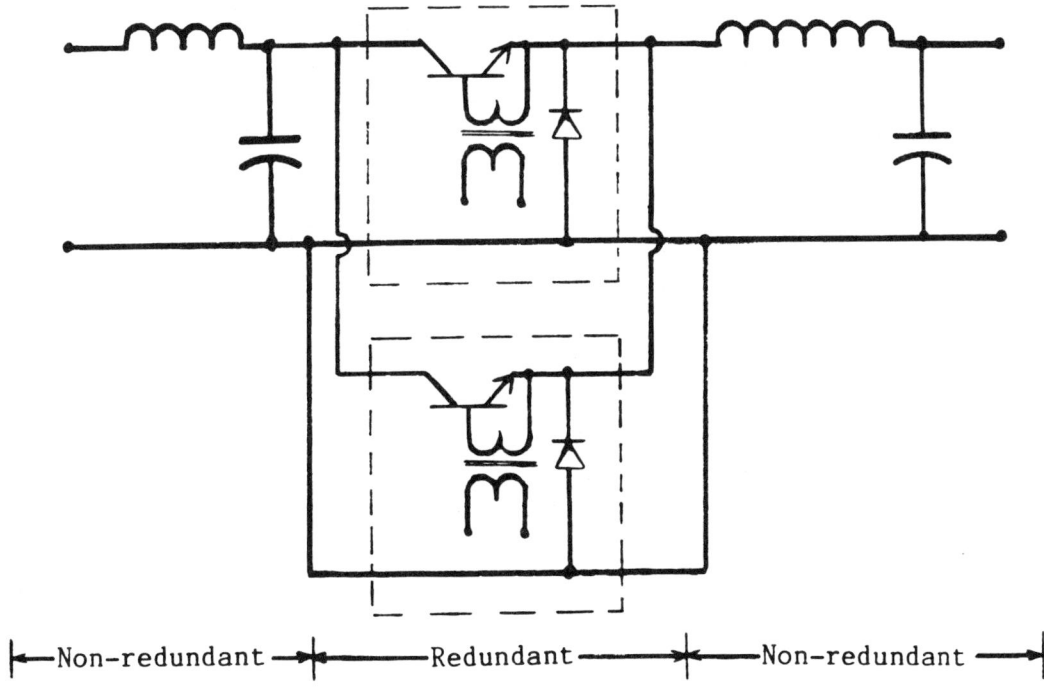

Fig. 5. Partial redundancy.

tinuously monitor power supply performance. Single or a particular combination of deviations in power supply performance parameters will be interpreted as a failure. Though different parameters, like temperature, frequency, current, or voltage at some point internal to the power supply may be used to detect a failure, input voltage (V_{in}), output voltage (V_{out}), input current (I_{in}), and output current (I_{out}) are normally chosen since these parameters avoid the dependency upon internal circuit configuration, operation, and are usually readily available. Also V_{in} and I_o need not be monitored as they depend mainly on source and load. Thus, it is sufficient to monitor the maximum value of I_{in} and the maximum and minimum values of V_o. The minimum value of I_{in} need not be monitored because V_o decreases if the source is not able to drive the load.

Conclusions

Various approaches to redundancy to improve the reliability of the systems have been presented considering power supplies as examples. One has to select the optimum redundancy approach depending upon various conditions giving proper weightages to cost, weight, size, complexity, etc. Need for a highly reliable failure detector is very essential.

References

1. J.K. Baker et al, "Power Conditioning Reliability Improvement through Standby Redundancy and Automatic Failure Detection," *IECEC Convention Record* 1967.
2. K.J. Jenson, "Synchronization and Failure Isolation for Redundancy Low Input Voltage Converters," *EASTCON Convention Record* 1967.
3. D.C. James et al, "Redundancy and the Detection of First Failures," *IRE Transactions on Reliability and Quality Control, Vol.RQC-11, No. 3, Oct 1962.*
4. *Edward J. Farrell, "Improving the Reliability of Digital Devices with Redundancy: an Application of Decision Theory" IRE Transactions on Reliability and Quality Control, Vol RQC-11, No.1, May 1962.*

RELIABILITY AND FAILURE MODE AND EFFECTS ANALYSES

1.0 Introduction

The importance of reliability has been illustrated and various methods and approaches to the improvement of reliability has been presented in the previous two papers. High performance, high levels of reliability and lower costs are the primary considerations in the design of any system. A reliability analysis provides a measure of inherent reliability. Thus, in this paper, the inherent reliability of a typical regulator is calculated as an example. Also included is the Failure Mode and Effects Analysis (FMEA) for a spacecraft power system. The primary purpose of FMEA is to identify and eliminate, where possible, critical single-point failures. Where critical single-point failures cannot be eliminated, the main goal is to reduce the probability of occurrence of such failures and to minimize their failure effects.

Thus a brief description of a spacecraft power system is presented. This facilitates easy understanding of reliability analysis and FMEA carried out in the following sections.

2.0 Description of a Spacecraft Power System

As a spacecraft power system is selected as an example to carry out FMEA, a brief description of the same is presented here to facilitate better understanding of FMEA. Fig. 1 shows a typical power system for a communication satellite in geosynchronous altitude. This power system is known as non-dissipative regulated bus power system as described in an earlier chapter entitled "Improved power conditioning unit for regulated bus spacecraft power system." A common control unit drives the power stages of shunt, charge, and discharge regulators. The charge regulator itself maintains the bus at a fixed voltage and charges the battery at a variable current. When the battery is completely charged, the shunt regulator is

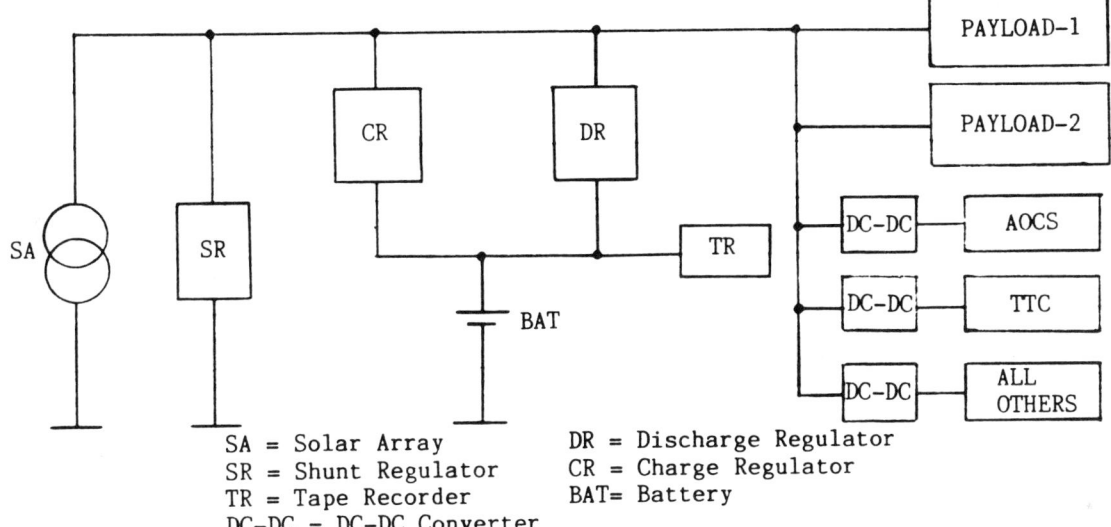

SA = Solar Array   DR = Discharge Regulator
SR = Shunt Regulator   CR = Charge Regulator
TR = Tape Recorder   BAT= Battery
DC-DC = DC-DC Converter

Fig. 1. Typical spacecraft power system (regulated bus).

turned-on and the charge regulator is switched from charge mode to trickle charge mode. Now the shunt regulator maintains the bus at a fixed voltage. During eclipse period, the battery discharge regulator, which is of boost type, boosts the battery voltage to bus level and maintains it at a fixed voltage.

The bus voltage is supplied to the payloads directly, whereas through dc-dc converter-regulators or dc-ac inverters to the bus loads, namely, the telecommand-telemetry-communication (TTC) system, attitude and orbit control system (AOCS), propulsion system, etc. Some of the loads like tape recorder, etc. are supplied from the storage battery through commandable redundant fuses.

2.1 Description of a Charge Regulator

As the charge regulator is selected as an example to show the reliability calculations, a brief description of the same is presented here. A detailed schematic of the charge regulator, and part of a spacecraft power system (Fig. 1) is shown in Fig. 2. The control is achieved by comparing the scaled-down bus voltage to a reference voltage. The error voltage is amplified, compensated, and fed to a pulse-width modulator through an attenuator. The modulator is synchronized at a clock frequency and its output is a pulse-width proportional to the

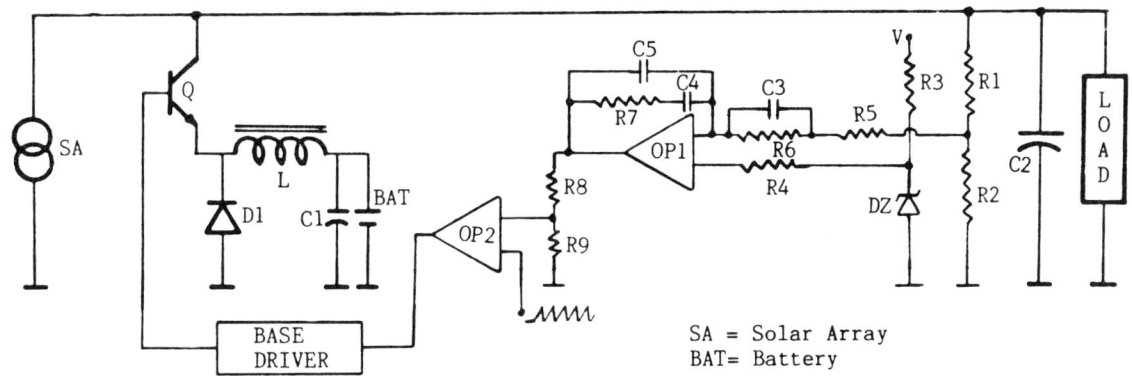

SA = Solar Array
BAT= Battery

Fig. 2. Detailed schematic of a charge regulator.

compensated-amplified error voltage but limited by the attenuator. This signal is employed to switch Q into saturation or cut-off. Thus, the current flow into the battery is controlled such that the bus voltage is maintained at V_o. When Q is on, energy is stored in the inductor (L) and this maintains a continuous current flow into the battery when Q is off.

3.0 Charge Regulator Reliability Calculations

To assess the inherent reliability of the charge regulator shown in Fig. 2, a reliability analysis is performed. The general assumptions are (a) all components are considered to be equally important for proper functioning, (b) the effect of one component failure on the other component is not considered. The MIL-SPECS of the components used in the calculations are given in Table 1 as extracted from MIL-HDBK-217A. The constant failure rate is calculated by

$$R = e^{-\lambda t}$$

where
- R = the reliability
- e = natural log base
- λ = failure rate
- t = duration in hours

Here a mission duration (t) of one year is assumed for reliability calculations. Standard reliability calculation techniques are used to calculate the reliability of different series and parallel components/circuits/units if there are any in the charge regulator. The reliability of the entire regulator is obtained by taking the reliabilities of all the components/circuits/units in series.

For series calculations

$$R_{series} = (R1)(R2)R3) \ldots (R_n)$$

For parallel calculations

$$R_{parallel} = [1-(1-R1)(1-R2)(1-R3) \ldots (1-R_n)]$$

where R1, R2, R3, ... R_n are the reliabilities of the components/circuits/units that are in series or parallel, respectively. Thus, the calculations along with the reliability calculated for a mission duration of one year is given in Table 2. The quality factors and the failure rates for various components of the charge regulator, as mentioned above, are given in Table 1. Table 2 shows five columns, namely, component description, failure rate, derating factor, number of components and effective failure rate. The failure rate shown in the second column in-

Table 1. Failure Rates of Components Employed in the Charge Regulator.

Components	Technology	Style	MIL-SPEC	Quality Factor	Failure Rate xE-9/Hr
Resistors:					
-Low Power	Carbon Composition	RCR	MIL-R-39008	S	0.45
-Low Power	Metal Film	RNR	MIL-R-55182	S	2.80
Capacitors:					
-Non-Polarized	Silver Mica	CMR	MIL-C-39001	S	0.30
-Polarized	Tantalum, solid	CSR	MIL-C-39003	S	11.00
Semiconductors:					
-Transistor	Silicon, NPN		MIL-S-19500	JANTXV	28.00
-Diode	Silicon		MIL-S-19500	JANTXV	12.00
-Zener			MIL-S-19500	JANTXV	17.00
-IC	Linear		MIL-M-38510	JAN CLASS A	12.00
Inductor:					
-Power Filter					7.50

NOTE:-Quality Factors, S = 0.01; JANTXV = 0.1; JANTX = 0.2; JAN = 1.0;

JAN CLASS A = 0.5; JAN CLASS B = 1.0

Table 2. Charge Regulator Reliability Calculations.

Component Description	Failure Rate	Derating Factor	Number of Components	Effective Failure Rate
RESISTORS				
Carbon				
Composition	0.0045	0.20	2	0.001800
Metal Film	0.0280	0.10	1	0.002800
	0.0280	0.20	5	0.028000
	0.0280	0.50	1	0.014000
CAPACITORS				
Tantalum, Solid	0.1100	0.30	2	0.066000
Silver Mica	0.0030	0.25	3	0.002250
SEMICONDUCTORS				
Diode	1.2000	0.50	1	0.600000
Zener	1.7000	0.20	1	0.340000
Transistor				
NPN, Silicon	2.8000	0.40	1	1.120000
IC (op)	6.0000	0.60	2	7.200000
IC (Driver)	6.0000	0.75	1	4.500000
INDUCTOR	7.5000	0.60	1	4.500000
Total Failure Rate				18.374850E-9/Hr

$$\text{Reliability, } R = e^{(-18.37485E-9)(8760)} = 0.999839$$

cludes the effect of quality factor. Derating factor is the ratio of the applied electrical stress to the rating of that particular component. The last column is the product of columns 2, 3, and 4. Charge regulator total failure rate is obtained by adding the last column, which is 18.374850E-9/hour. Finally, the probability of the charge regulator functioning as per the design meeting expected performance over a period of one year is 0.999839.

4.0 Failure Mode and Effects Analysis

As mentioned above, the primary purpose of FMEA is to identify and eliminate, where possible, critical single-point failures. However, if critical single-point failures cannot be eliminated, the main goal is at least to reduce the probability of occurrance of such failure and to minimize their failure effects. Critical single-point failures are those failures occurring singly that disable the power system from providing power to spacecraft critical loads.

To realize the complete mission goal, all the subunits are essential. One can equally envisage the cases where certain equipment fails without affecting the complete system and hence can partially fulfill the mission goals. FMEA is carried out on subsystem/unit level and for each unit analyzed, all failure modes are included. Structure and passive elements such as power busses and wiring are excluded from the analysis. The basic failure modes considered are open, short, and degradation.

Thus, the FMEA carried out on spacecraft power system is presented in Table 3. From this table, it is clear that to avoid single-point failures, the charge, discharge, and shunt regulators have to be redundant. The storage battery shall have cell open failure bypass circuitry and shall be over designed with respect to the storage capacity to avoid any effect due to cell shorts. In addition to connecting the tape recorder to the battery directly, some switching mechanism shall be incorporated such that it can also be powered from the regulated bus.

Conclusions

The importance of reliability and various approaches to the improvement of reliability has been presented in the previous two papers. In this paper, the inherent reliability of a typical regulator is calculated as an example. Also included is the Failure Mode and Effects Analysis for a spacecraft power system. FMEA helps to identify and eliminate critical single-point failures.

Table 3. Spacecraft Power System Failure Mode and Effects Analysis.

Subsystem/Unit	Failure Mode	Effects on other systems
Solar Array Section	Open	As solar array consists of n sections and are connected using isolation diodes, opening of a section only reduces the solar array output by 1/n.
	Short or degradation	Same as above.
Shunt Regulator	Open	Bus voltage will not be maintained. Loads connected to the bus may not accept the variation in the bus voltage, which is the solar array voltage.
	Short	Shorts the solar array and some component may fuse open. Bus voltage will not be maintained.
Charge Regulator	Open or degradation	Battery cannot be charged. Effect is the same as the loss of the battery.
	Short	Bus voltage clamps to battery voltage. Battery charge rate is not controlled. Once the battery is charged up, the battery has to be protected by opening the battery. Whenever the battery is charged, the bus loads cannot work. However time sharing is possible.
Discharge Regulator	Open	Battery cannot be discharged. Loads cannot be supplied power during eclipse period.
	Short or degradation	Battery voltage cannot be boosted to bus voltage and bus voltage clamps to battery voltage. Hence, battery has to be opened.
Battery	Open	Tape recorder cannot operate. Eclipse operation of the satellite is not possible.
	Short	Power will be shunted to ground. Battery has to be opened. Taperecorder cannot operate. Eclipse operation of the satellite is not possible.
DC-DC Converters	Open	Automatically changes to the redundant unit.
	Short	Automatically changes to the redundant unit.

References

1. P.R.K. Chetty, *"Spacecraft Power Systems—Some New Techniques for Performance Improvement,"* Ph.D. Thesis, Indian Institute of Science (I.I.Sc), India, 1978.

Index

Index

A

assembly
　fabrication, 166

B

base drive, 101
　proportional, 126
batteries, 147
battery capacity, 150
battery charge-discharge control, 149
boost converter
　modeling of, 29, 36
boost converter modeling, 12, 19
boost regulator, 5
buck regulator, 5
buck and buck-boost converters modeling, 14

C

CAD package, 84
capacitors, 102
cell array
　solar, 146
cells
　solar, 146
　storage, 147, 148
charge mode operation, 154
charge regulator
　description of a, 171
charge regulator reliability calculations, 172
CIC mode, 8, 15, 27
CIECA, 8, 9, 15, 17, 18, 33, 39
　review of, 26
circuit design
　reliable, 160
circuits
　interface, 113
　relay driver, 163
　TTL integrated, 164
closed loop frequency response measurements, 76
closed loops, 60
closed-loop configuration, 60
common regulator, 161
comparators, 108
compensation, 103
compensation network, 58, 61, 62, 63
component reliability, 165
component selection, 101
component tolerances, 165
conditioning
　power, 148
control IC, 102, 132
　description of, 127
　special purpose, 133
control ICs for switch mode power supplies, 132, 134, 135, 136

control system, 160
converter
　basic, 49
　boost, 29, 49, 72
　buck, 23, 49, 72
　buck-boost, 15, 24, 49, 72
　Cuk, viii, 40, 45
　dc-dc, 123
　modeling, 52
　modeling of boost, 36
　modeling of stabilized current programmed boost, 30
　pulse-width modulated push-pull, 96
　push-pull, 95
　switching dc-dc, 48
converter modeling
　boost, 12
converter operation, 124
converter-modeling
　boost, 19
converters
　buck and buck-boost, 14
　buck, 21, 38
　buck-boost, 21, 38
　current programmed switching dc-dc, 33
　dc-to-dc, v
　instability in current programmed, 28

177

modeling of, 72
modeling of stabilized current programmed buck and buck-boost, 32
switching, 17
switching dc-dc, 8, 51, 70
switching dc-to-dc in the CIC mode, 27
converters dc-to-dc, vi, 1
core material selection, 99
Cuk converter, 40, 45
modeling of, 42
Cuk converter, viii, 1
current programmed converters instability in, 28
current programmed switching dc-dc converters, 33

D

dc-dc converter, v, vi, 123
decoding scheme, 113
design
solar cell array, 150
design considerations, 149
design reliability, 160
detector
failure, 169
DIC mode, 15, 38, 39
digital interface, 163
digital shunt regulator, 106, 118
diode
snubber design, 100
diodes, 102
discharge mode operation, 155
dissipative analog shunt regulator, 108
drive source interfaces, 162
drive unit
solar array, 149
driver ICs, 137
driving, 161

E

electronics
power, 2
power processing, 2
EMI, 6
energy sources, 146
error amplifier and compensation, 53

F

fabrication reliability, 165
fabrication/assembly, 166
failure detector, 169
failure mode, 173
failure mode and effects analysis, 173
filter damping network, 91
filter design, 99
filtering
input, 161

free-running switch-mode power supplies, 133
frequency limitations, 111
frequency response measurements, 74
closed loop, 76
open loop, 74
full-wave rectification, 98

G

gain
low frequency, 85

H

hardware implementation, 111

I

IC
control, 102
timer, 141
IC timers as controllers for switch-mode power supplies, 140
ICs
driver, 137
microprocessor switch-mode power supply control, 137
impedance
input, 87
output, 85
implementation
hardware, 111
inductor design, 99
injection
signal, 77
injections
magnetic, 67
input filtering, 161
input impedance, 87
inputs driven from switches, 161
interface
digital, 163
interface circuits, 113
interface with relays, 163
interfaces
drive source, 162
interfacing, 112

L

level shifting, 164
limitations
frequency, 111
line transmission, 88
line transmission characteristic, 88
linear power supplies, 127
linear power systems, 3
load step response
output, 89
logic redundancy
majority, 168
loop gain, 69

loop gain and phase determination, 76
loops
closed, 60
low earth orbit vs geosynchronous orbit, 147
low frequency gain, 85

M

magnetic injections, 67
mean time between failure, 2
microprocessor controller, 109
microprocessor switch-mode power supply control ICs, 137
microprocessor-controlled digital shunt regulator, 105
mode
CIC, 8
effects analyses, 170
failure, 170, 173
reliability, 170
modeling a switching-mode power supply, 64
modeling of Cuk converter, 42
modulator
pulsewidth, 54
MTBF, 2
multiphase operation of self-oscillating switching regulator, 119

N

narrow band tracking voltmeter
necessity for, 74
network
compensation, 58
filter damping, 91
phase-shift, 121
nondissipative, 5

O

off-line switcher, 94
design of, 97
open loop
driving an, 66
open loop frequency response measurements, 74
operation
charge mode, 154
discharge mode, 155
shunt mode, 153
optimization, 169
optoisolator, 95, 103
orbit
geosynchronous, 147
low earth, 147
output
pulse width modulated, 95
output impedance, 85
output load step response, 89

P

phase determination, 76

phase-shift network
  description of, 121
point of injection, 76
power budget, 149
power conditioning, 160
power conditioning and control system, 148, 160
power conditioning unit, 151, 152, 156
  theory of improved, 153
power electronics, v, 2
power flow control logic, 156
power processing electronics, 2
power stage, 48
power supplies
  improvements to, 128
  linear, 127
  switch-mode, v, vii, 2
power supply, 2
  switching-mode, 60, 64
power switching transistor, 101
power system
  regulated bus for, 151, 152
power systems
  dissipative, 3
  linear, 3
  nondissipative, 3, 5
  spacecraft, 146
pulse-width modulated output, 95
pulse-width modulated push-pull converter, 96
pulse-width modulator, 54
push-pull converter power stage, 95
PWM, 3
PWM push-pull converter, 94
PWM switching regulator, 123

R

rectification, 98
  full-wave, 98
redundancy, 166, 167
  load sharing, 167
  majority, 168
  majority logic, 167
  partial, 169
  shared mode of, 168
  shared mode of standby, 167
  standby, 167
redundancy approaches, 167
regulated bus power system, 152
regulator
  boost, 5
  buck, 5
  buck-boost, 5, 142
  charge, 171, 172
  digital shunt, 105, 106, 118
  dissipative analog shunt, 108
  microprocessor-controlled digital shunt, 105
  PWM switching, 123
  self-oscillating switching, 119

series, 3
shunt, 4
switching, 73
two phase self-oscillating switching, 122
regulator switching, 69
regulators
  switching, 48
  switching free-running, 6
relay driver circuits, 163
relays
  interface with, 163
reliability, 167
reliability and redundancy, 167
reliable circuit design, 160

S

self-oscillating switching regulator
  multiphase operation of, 119
series regulator, 3
  a basic configuration of, 4
shifting
  level, 164
shunt mode operation, 153
shunt regulator, 4
  a basic configuration of, 4
signal injection, 77
simulation, 84
simulator
  solar cell array, 107
SMPS, 132
snubber design, 100
snubber diode design, 100
software, 112
solar array drive unit, 149
solar cell array, 146
solar cell array area, 149
solar cell array design, 150
solar cell array simulator, 107
solar cells, 146
solar power systems, 118
spacecraft power system
  building blocks of a, 146
  description of a, 170
  regulated bus, 151
spacecraft power systems, 146
special purpose control ICs, 133
stability definitions, 56
stabilized current programmed boost converter
  modeling of, 30
stabilized current programmed buck and buck-boost converters
  modeling of, 32
standby redundancy
  shared mode of, 168
storage cells and batteries, 147
storage cell types, 148
switch mode power supplies
  control ICs for, 132
switch timing, 109

switch-mode power supplies, 2, 60
  control ICs for, 134, 135, 136
  driven type, 134, 135, 136
  free-running, 133, 138
  IC timers as controllers for, 140
switch-mode power supplies v, vii
switcher
  off-line, 94
switches
  inputs driven from, 161
switching converters, 17
switching dc-dc converter, 8, 48, 51
  modeling of, 70
switching dc-to-dc converters in the CIC mode, 27
switching regulator
  typical, 48, 49, 73
switching regulator controller
  timer IC as, 141
switching regulator transfer functions, 69
switching regulators, 48
  CAD package for the design of, 84
  free-running, 6
  modeling and design of, 48
system considerations, 149
systems
  control, 148

T

testing, 165
timer IC
  description of, 140
timer IC as switching regulator controller, 141
transfer functions
  compensation network, 61, 62, 63
  measurement of, 79
transformer, 97
transformer winding technique, 98
transistor
  power switching, 101
triangles
  vector, 78
TTL integrated circuits, 164
two phase self-oscillating switching regulator, 122

V

vector triangles, 78
voltage selection, 149
voltmeter
  narrow band tracking, 74

W

winding technique
  transformer, 98